continued on back

Mathematics, Statistics, and Systems for Health

Mathematics, Statistics, and Systems for Health

NORMAN T. J. BAILEY

Unit of Health Statistical Methodology
World Health Organization

JOHN WILEY & SONS

Chichester · New York · Brisbane · Toronto

Library of Congress Cataloging in Publication Data:

Bailey, Norman T. J.
Mathematics, statistics, and systems for health.

(Wiley series in probability and mathematical statistics)

Based on the author's The mathematical approach to biology and medicine, published in 1967.
Includes bibliographical references and indexes.
1. Medicine—Mathematics. 2. Medical statistics.
3. System analysis. 4. Operations research.
I. Title. [DNLM: 1. Biometry. 2. Mathematics.
QH323.4 B155m]
R853.M3B33 610'.1'51 77–1307

ISBN 0 471 99500 2

Photosetting by Thomson Press (India) Limited, New Delhi
Printed by The Pitman Press, Bath

‘4.116 Alles, was überhaupt gedacht werden kann, kann klar gedacht werden. Alles, was sich aussprechen läßt, läßt sich klar aussprechen. . . .

7 Wovon man nicht sprechen kann, darüber muß man schweigen.’

LUDWIG WITTGENSTEIN
Logisch–Philosophische Abhandlung

Preface

The burden of disease and ill-health in the world today is truly staggering. In the developing countries the almost universal occurrence of chronic infectious and parasitic disease, as well as serious malnutrition, holds back the economic and social advances needed to overcome these diseases. In developed countries there is widespread anxiety about many noncommunicable diseases, especially cancer and cardiovascular disease, and the incidence is actually increasing in some cases. Moreover, the costs of health services in advanced countries have soared with continually increasing demands; while in vast underprivileged areas of the world, health care is virtually non-existent.

The purpose of this book is to make available to a wide reading public a greater understanding of the role of quantitative methods in helping to develop and improve medical care and health services throughout the world. Special attention has been paid to the significance and potentialities of mathematics, statistics, modelling, computers, operational research, systems analysis, and system dynamics. All of these subjects, which have, I believe, a major part to play in the promotion of health, are liable to entail a considerable degree of technical complexity. They are, moreover, targets for much current debate and criticism.

Fortunately, a general understanding and appreciation of the basic ideas and implications of these techniques can, in my opinion, be derived from straightforward commonsense notions that are familiar to almost everybody. This means that the principles of the methodologies mentioned, together with their implications, can be made readily accessible, not only to health professionals who could profit from them, but also to those members of the lay public who are concerned with the consequences for their daily lives. A firmer basis can be supplied for much-needed multidisciplinary discussion amongst experts, including of course health professionals and mathematicians. Assistance can also be given to the development of a wider participation of the general public in the planning, implementation, and evaluation of medical and health services. Health is too important a subject for the control of mathematically based support to be left entirely in the hands of mathematical experts—indeed, this could be positively dangerous. Hence the need for basic ideas and principles to be put in simple terms and widely understood.

These objectives are ambitious, and no doubt only very imperfectly realized in the present book. Nevertheless, the potential advantages of the various mathematical approaches discussed are very great, and, if generally understood, could make major contributions to the solution of crucial health problems in the world at the present time.

It should be mentioned at this point that the book is largely based on a previous publication, *The Mathematical Approach to Biology and Medicine*. This was originally published in 1967, but has been adapted and extended to throw the major emphasis onto applications to medicine and public health. Also, the first edition was in two parts, the first of which comprised a rather general discussion of principles and techniques, while the second part dealt with a number of highly specific and mathematically sophisticated applications. The new edition, on the other hand, adopts an integrated treatment, and keeps the purely mathematical aspects down to a minimum. It is thus more concerned with general notions and principles than with technical mathematical details. I hope that this mode of presentation will be helpful in promoting the objectives described above.

I should like to express my indebtedness to the Publisher for making this new version and orientation of the original book possible. I should also like to take this opportunity of thanking Mrs Mary Martin and Mrs Elise Juni for their assistance in preparing the revised typescript. And finally, for invaluable assistance with the proof-reading, I must thank Mrs Mary Martin, Miss Susan Hackshall, and my wife, Mrs Jean Bailey.

Geneva
June 1977

NORMAN T. J. BAILEY

Contents

CHAPTER 1

Human Health Problems

1.1 DISEASE AND ILL-HEALTH IN THE WORLD

The traditional scourges of history, namely, war, pestilence, and famine, are unfortunately still very much with us. Although global war has been absent for some thirty years, this ultimate disaster has been replaced by a continuous series of localized wars, while the general increase of violence in society is a matter for world-wide concern.

The tragic role of pestilence, in the form of infectious or communicable disease, has in the past had far greater consequences for human misery and suffering than even the most ferocious wars. Thus the Black Death in the 14th century caused possibly 25 million deaths out of a total population of 100 million in Europe. Local death-rates were even higher, with whole villages or towns being practically wiped out. In 1520 the Aztecs lost about half their population from smallpox. The Russian typhus epidemics during the years from 1918 to 1921 involved around 25 million cases with a death-rate of ten per cent. And the world influenza pandemic of 1919 is believed to have had a death role in the region of 20 million.

Smallpox is now happily almost extinct, and may soon disappear altogether. Tuberculosis has steadily dropped in prevalence, but there are still about 15 million active cases in the world. Syphilis, the Great Pox, no longer has the virulence or prevalence of earlier times. Nevertheless, venereal disease trends, especially gonorrhoea, are rising and causing widespread concern. In developed countries at any rate, measles is well under control, but in Africa for example it is still a serious cause of mortality. And who can say whether a new virulent strain of influenza may not suddenly arise, leading to yet another global pandemic?

Although we may expect that modern methods of public health control will be able to avert the worst disasters arising from purely epidemic outbreaks, there are many diseases, especially those of a parasitic nature, that are still endemic on an astronomical scale. Thus there are still approximately 350 million people in the world living in areas where malaria is endemic, and where no organized antimalarial programs have yet been established. Moreover, the prevalence of malaria is increasing in many places. Again, schistosomiasis is believed to affect around 200 million people, with a further 600 million being exposed to infection. The figure for hookworm disease is 450 million. Trachoma, for which methods of treatment have been improving, may involve a total of 400–500 million. This is only part of the communicable disease picture in the world today. The total load of human suffering is truly staggering, and represents a formidable challenge to medical science and public health.

In addition to the above we have to consider the increasing role of noncommunicable disease. Here the effects are most apparent in the developed countries, but even now the consequences are beginning to be felt in some developing countries as standards of living improve.

Thus the heavy toll taken by cardiovascular disease and cancer is well-known. It has been estimated that in Europe and America approximately 15–20 per cent of the newborn will, at current mortality rates, ultimately die of malignant neoplasms. In Europe round about 50 per cent of the newborn will become victims of cardiovascular disease, and in America the figure will be even higher, approximately 60 per cent. Much of this disease occurs today in age groups where people would otherwise be leading active lives. The mortality and morbidity that occur therefore entail serious restrictions in personal expression and development as well as major losses to society as a whole.

These noncommunicable diseases are of course only the highlights. There is a wide range of other conditions involving mental ill-health, dental disease, road accidents, occupational hazards, etc., which all add up to a considerable burden of mortality and morbidity.

Further, there are the problems of malnutrition. A world-wide survey has suggested that in the developing countries there are at least 100 million children under 5 years of age suffering from moderate to severe forms of protein–calorie deficiency. Such conditions are liable to lead to physical and mental handicaps as well as lowering resistance to infectious disease. Moreover, recent years have witnessed a reversal in previous trends towards an improvement of nutritional status. For large, already underprivileged, segments of mankind there has been a deterioration in food supplies, and the quality of life is being degraded even further. In addition, nutritional anaemias, particularly iron-deficiency anaemias, are widespread. Even in the USA it is estimated that 20 per cent of the population is affected. In Europe, the figure for women varies between 10 and 25 per cent. Even higher rates have been recorded in some developing countries. Widespread chronic undernutrition is evidently a major threat to world health.

While most diseases are strongly influenced by environmental factors, the biological mechanisms of infection and spread in the communicable diseases are usually simpler to identify. This facilitates ultimate control by direct attack on the infectious organisms themselves, or by indirect public health methods such as inoculation. The noncommunicable diseases, on the other hand, tend to be influenced by complicated patterns of multiple causation. This makes control much more difficult. Even when a major link can be identified we may be unable to use the knowledge to reverse a rising trend of incidence, as for example with the association between smoking and lung cancer.

Hippocrates, in his *Airs, Waters and Places*, appears to have been well aware of the general importance of environmental factors. But in recent decades there has been a continually increasing understanding in detail of the role of the environment in disease causation, and the urgent necessity of promoting and maintaining good ecological conditions is becoming widely accepted. More and more people are realizing the obvious degradation of the quality of life resulting from a wide variety of environmental pollutions. Many individuals can learn to tolerate noise, but it can inflict permanent damage on the hearing. And certain kinds of noise are becoming increasingly unacceptable, as indicated by the widespread objections to the close proximity of airports and the specific irritations of sonic booms.

The importance of clean water supplies is fairly generally understood. Moreover, the consequences of the growing pollution of lakes and rivers with the obvious destruction of plant and fish life are clear for all to see. Equally obvious are the unpleasant effects of air pollution. In spite of clean-air programs in many cities, a general tendency for smog to increase in urban and industrial areas can be observed directly in most parts of the world.

Again, the health hazards of atomic fall-out and nuclear waste disposal give rise to further major problems closely related to the desperate need to develop new sources of power. Not only are the traditional fossil fuel supplies used by advanced countries dwindling, but vast areas of population in developing countries are beginning to demand the standards of living achieved in the more developed countries. And to achieve any substantial progress in this direction requires industrialization and a great increase in supplies of energy.

While many of these new problems are not specifically in the health area itself, there are direct threats to health arising from immediate environmental factors, especially the multifarious forms of pollution mentioned, as well as from the more intangible consequences of interfering with well-balanced ecological systems that have been established over very long periods of time. Well-intended efforts to use new methods of technological control may all too easily turn out to be counter-productive. For example, the introduction of piped water supplies in African villages without adequate drainage may increase the incidence of malaria because the additional pools of water provide new breeding sites for mosquitoes. Similarly, the construction of dams and man-made lakes for irrigation may be offset by an increase of schistosomiasis spread by the growth of new snail populations, and by the rapid rise of water tables in some areas with consequent salinization of the soil and crop destruction.

This sketchy account of some of the major aspects of the world's health and disease predicament is simply intended to indicate the magnitude and range of the problems involved. For authoritative discussion of details it is of course necessary to consult the professional medical and public health literature. But, in the context of the present book, it is sufficient to point out the scale of the unsolved problems, and to suggest that perhaps a better understanding of quantitative methods and a more systematic approach to the information available could at least assist in achieving some marked improvement in a horrendous situation.

1.2 WHAT IS HEALTH?

Having taken a brief look at the scale of some of the world's major health problems, we might well ask for a working definition of health as a guide to the scope and objectives of subsequent discussions.

The Constitution of the World Health Organization, supported at the time of writing by some 151 member states, is introduced by a statement of basic principles to be used for the purposes of cooperation between the contracting parties, and the promotion and protection of the health of all peoples. The first three of the nine principles enunciated declare that:

> Health is a state of complete physical, mental and social well-being and not merely the absence of disease or infirmity.
>
> The enjoyment of the highest attainable standard of health is one of the fundamental rights of every human being without distinction of race, religion, political belief, economic or social condition.
>
> The health of all peoples is fundamental to the attainment of peace and security and is dependent upon the fullest cooperation of individuals and States.

The initial statement of principles is then followed, reasonably enough, by Article 1 which affirms that: 'The objective of the World Health Organization... shall be the attainment by all peoples of the highest possible level of health.'

While these noble aspirations certainly help to provide a basis for political, administrative, and technical action, there is a truly appalling gap between the stirring ideals and the grim realities of Section 1.1. Medical and health workers the world over, in private practice, hospitals, research institutes, universities, regional administration, national ministries of health, and international organizations, are all struggling to put into practice the general principles that most of us support. However, there are to begin with almost insuperable difficulties caused by lack of knowledge and shortages of resources, in terms of time, money, equipment, staff, etc., not to mention the relative inaccessibility of many of those whose needs are greatest. These practical constraints lead to a whole set of additional problems, which are not specifically medical but which involve rather questions of logistics, management, organization, and planning. More will be said about these later, including the possible use of suitable quantitative methods to facilitate appropriate action.

1.3 WHO IS HEALTH FOR?

Who is health for? Health is for all people, as indicated in the WHO Constitution referred to in the previous section. However, this is a *de jure* statement, and we may well ask, what happens *de facto?*

Ivan Illich (1975) closes his recent, and highly controversial book, *Medical Nemesis* with the view that,

> Man's consciously lived fragility, individuality and relatedness make the experience of pain, of sickness and of death an integral part of his life. The ability to cope with this trio autonomously is fundamental to his health. As he becomes dependent on the management of his intimacy, he renounces his autonomy and his health *must* decline. The true miracle of modern medicine is diabolical. It consists not only of making individuals but whole populations survive on inhumanly low levels of personal health. That health should decline with increasing health service delivery is unforeseen only by the health managers, precisely because their strategies are the result of their blindness to the inalienability of life.

Not everyone agrees with this statement, and many are fundamentally opposed to it. Nevertheless, it does represent a view that is becoming increasingly acceptable, and as such represents a challenge that cannot be ignored. It is abundantly evident that in the world today, especially in developed countries, enormous sums of money are being invested in medical training, medical technology, and medical research. Well-organized efforts are made to market a deluge of new pharmaceutical preparations. Health planning is advocated on an ever-increasing scale. Yet global standards of health and well-being *are*, if anything, declining. And, as we have seen in Section 1.1, malnutrition is widespread, parasitic diseases are rampant and malaria is even increasing, many cancer rates are rising, and so on.

When the British National Health Service was originally planned by William Beveridge, it was assumed that there existed in the community a 'limited quantity of morbidity'. Sickness-rates would therefore fall as treatment became more readily available, and the annual cost of the Health Service would correspondingly decline. But, in fact, demands for treatment have continually increased, and costs have soared.

This seems to be a world-wide phenomenon. It derives partly from psychological causes in the individual, but is also aggravated by the unfettered development of the health service industry. The latter is not only the largest *service* industry in the world, but is also one of the largest industries in absolute terms in most countries. In the USA, for example, it ranks second, behind the food industry and ahead of defence.

The massive industrialization of health production is of course geared towards maximizing the profits and defending the vested interests of the producers. It is supported by the formation of defensive cartels and large-scale advertising, both to doctors and members of the public. This means that the ultimate consumer of the health services, i.e. the individual sufferer or patient, is generally rather poorly informed about what health facilities exist and how to make a critical assessment of them. He is, moreover, usually in a disadvantageous position *vis-à-vis* the overwhelming psychological and social pressures generated by vested interests.

All this is part of the process that Illich calls the medicalization of health. He sees an increasing degree of *iatrogenesis* in society, which is *clinical* when it results from medical care itself; *social* when it is generated by ill-conceived health policies and industrial organization; and *structural* when the capacity of individuals for mutual self-help is seriously undermined by medically sponsored attitudes and behaviour.

Health and the benefits of medical science are, then, *for people*. They are not, primarily, for the medical profession, medical research workers, health industries, health planners, ministries of health, etc., although all of these may have essential parts to play in achieving the overall objective.

1.4 WHAT CAN BE DONE?

The yawning chasm between the ideal of 'complete physical, mental, and social well-being' for all peoples and the existing overall level of achievement in the world has already been emphasized in Section 1.2. Medical workers and health authorities are everywhere putting huge efforts into delivering better health services. Unfortunately, not only are many of the essential resources extremely scarce, but there are powerful social forces, as indicated in Section 1.3, that tend to produce very one-sided and inequitable developments.

Although this book is naturally not about curative and preventive medicine, or public health, as such, it *is* about the assistance that can be given to these areas of endeavour by appropriate quantitative methods. As we shall see in more detail later, the types of mathematical, statistical, or computing support that may be useful are very much conditioned by the overall aims and objectives of society in the health field. Or, to put it in more concrete terms, by the aspirations of people with regard to the levels of health and relative absence of disease that are feasible.

This question of feasibility, and indeed of desirability, raises many crucial issues that are becoming more and more widely appreciated in the world today. If medical care is in short supply, who has or should have the highest priority for treatment? When slender financial support is available in a developing country for some expansion of health activities, how much should go, for example, towards building up a centralized acute general hospital to serve perhaps a few hundred or thousand people a year, compared with primary rural health services urgently needed by hundreds of thousands, if not millions, of people? Again, how much money should be devoted to highly specialized techniques such as heart transplantation, renal dialysis or neurosurgery, as compared with vaccines to protect whole communities? Such examples could be multiplied indefinitely.

The whole problem of priorities and constraints is in fact present at nearly all levels of enquiry. Even a purely research investigation, requiring the statistical planning of an epidemiological survey or laboratory experiment needing a special statistical design, will be conditioned by the kind of questions that are of major interest, the precision thought to be necessary, the staff, time, and finance available, the deadlines to be met, and so on. Such considerations are even more acute when one is concerned with practical applications in operational contexts.

It is abundantly clear that extensive medical knowledge plus the mechanisms for translating this knowledge into the appropriate health services are needed on a very wide scale. However, whatever is undertaken should ultimately, in my view, be directed towards what the consumers of health care really need. That is, activities that are merely profitable to industry or merely interesting to doctors and research workers should not be undertaken at all, or at least should be given a low priority.

A properly balanced supply of doctors, nurses, auxiliaries, drugs, diagnostic and therapeutic aids, with adequate facilities for distribution, plus realistic and practically oriented programs of research, are therefore all essential. Moreover, there are still wider issues involved in the maintenance of health. Other sectors are involved, such as education, agriculture, water resources, energy production, environmental factors, not to mention economics. Somehow, methods have to be found for manipulating this whole complex system of interacting activities, so as to provide the best reasonable level of health and well-being that can be achieved in relation to other urgent demands.

Regional and national health planning authorities have to work towards improving the existing situation, so that all those who in some sense represent the interests of the health consumers can help to achieve what is potentially feasible. This is, of course, a formidable task, requiring enthusiastic and intelligent cooperation between the medical professions, medical technology, pharmaceutical industries, health administrators, and planners, as well as the whole community of health consumers themselves. Moreover, the successful development of such an enterprise can only be achieved if proper allowance is made for the inevitable constraints of society, whether these be religious, psychological, moral, political, or economic.

1.5 THE ROLE OF QUANTITATIVE METHODS

The foregoing discussion, which is no doubt somewhat biased and personal, has at any rate emphasized the depressing magnitude of the world's medical and health problems, which still exist in spite of the continually increasing range of medical knowledge and skill. The need for more extensive understanding is evident. Equally evident is the need to apply that understanding on a scale hitherto found impossible.

Such is the background to the present book, which is about the help that can be given to improving methodological approaches to the spectrum of problems arising. Our thinking needs to be clarified. A more scientific approach to many difficulties might prevent much useless effort. Theories should be clearly formulated and properly tested in practice. The real needs and wishes of the ultimate consumers of health, people themselves, must be adequately met. The implications of research must be evaluated. The probable consequences of administrative decision making and planning processes must be intelligently appraised. Social, political, and other constraints must be suitably allowed for. Better ways must be found of understanding and controlling the complicated and interactive systems that inevitably arise. And so on.

All these requirements are proper subjects for the use of appropriate quantitative methods: mathematics, statistics, computers, operational research, systems analysis, etc. These technical aids are themselves, like medicine and public health, foci for investigation and development by specialists. On the other hand, if they are to be accepted and reliably used in applications, the general assumptions, concepts, and procedures must be widely understood and appreciated in general nonspecialist terms.

The extent to which quantitative approaches and methods can help us to grasp the thorny problems raised by all the wider issues that have been mentioned is a matter of hotly contested current debate. Mathematical analysis is well established in biochemical and physiological research, good statistical design is seen as essential to well-executed epidemiological surveys or randomized field trials, and computer methods are accepted as sensible technological aids to handling large quantities of basic data. But the role of operational research and systems analysis in clarifying the behaviour of large complexes of interacting components, where well-meaning decisions can so easily be counter-productive, is still a matter for argument and further investigation.

The following chapters therefore attempt to set out in relatively straightforward language what these quantitative methods are all about and what the implications are. It is my conviction that a wide range of mathematically based concepts and skills can, for purposes of application and evaluation, be appreciated in general terms and discussed in a nontechnical way with those experts whose job it is to handle the technical details. This book is, accordingly, not intended as a text to teach mathematical methods. It aims, instead, at enabling health workers to grasp the principles involved and to help them decide more readily whether such approaches might be usefully called upon to solve some of their own problems.

FROM THE DESK OF

J. STUART HUNTER

Bob:

Here's that book I mentioned earlier today. Your friend is certainly free to keep it a few months, if needed.

Stu

CHAPTER 2

The Scientific Approach

2.1 THE SCIENTIFIC METHOD

'The unity of all science', says Karl Pearson (1892) in *The Grammar of Science*, 'consists alone in its method, not in its material.' Although we are primarily concerned with applications to the field of medicine and health, it is worth giving some preliminary thought to the general principles involved in scientific research and investigation of any kind. We shall then be able to see more clearly the implications for success or failure in our own special areas of application.

Broadly speaking, the scientific method consists of a continual process of testing, modifying, and developing ideas and theories in accordance with the dictates of the available evidence. In a sense it is really no more than an extension of ordinary common sense and rational behaviour. At the same time, it is true that many highly developed pieces of scientific research are so complex and specialized, and couched in such obscure jargon, that their relevance to everyday life is not immediately obvious. Nevertheless, the requirement that everything must ultimately be interpretable in factual terms means that the general implications can always be explained, at least approximately, in commonsense language.

In principle, Wittgenstein's propositions No. 4.116 and the final No. 7 from his *Logisch–Philosophische Abhandlung* (1921) are highly relevant, namely, 'Alles, was überhaupt gedacht werden kann, kann klar gedacht werden. Alles, was sich aussprechen läßt, läßt sich klar aussprechen,' and 'Wovon man nicht sprechen kann, darüber muß man schweigen,' In other words, what can be thought at all can be thought clearly; what can be expressed can be expressed clearly; and what we cannot say we must keep quiet about. These salutary notions, if consistently applied, would not only reduce the existing mountains of scientific and pseudo-scientific verbiage to more manageable proportions, but would also make the significant content far more intelligible and far more widely appreciated.

Some of the most striking examples of the scientific method come from the exact sciences of physics and chemistry, though we do not want to get involved in technicalities that are too far removed from the special field under discussion. However, in point of fact, some highly impressive illustrations of the development of the scientific method as an operational instrument occur in the history of astronomy (see also, Section 3.1). The gradual emergence and acceptance of the modern picture of the solar system is widely and brilliantly portrayed in Arthur Koestler's *The Sleepwalkers* (1959). This is eminently worth reading for the insight it displays into the interactions between the purely intellectual, philosophical aspects of the scientific method and all the constraints and conditioning imposed by human factors and technological advances.

The Greek astronomer, Aristarchus of Samos (*c.* 310–230 BC), already held the heliocentric view. This was known to his younger contemporary, Archimedes, but no further developments took place. Instead, the authority of Plato and his predilection for perfect spheres and uniform motions in perfect circles led in a different direction. Astronomers kept the earth at the centre of the universe and tried to explain the apparent motion of the sun, moon and five planets in terms of combinations of spherical motions. Aristotle described a system that entailed a total of fifty-four spheres, suitably nested and interrelated. Later refinements culminated in the tightly organized maze of epicycles devised by Ptolemy of Alexandria in the second century AD. Some forty of these 'wheels within wheels' were required, constituting what Koestler has aptly called Ptolemy's Ferris Wheel Universe.

Such ideas held the ascendancy for the best part of the next millenium and a half, although heliocentricism was never entirely forgotten. However, it was not until the firm rejection of the geocentric universe by Copernicus that the tide began to turn. In his book, *De revolutionibus orbium coelestium*, published in 1543, the year of his death, Copernicus clearly had the sun in the centre with the planets, including the earth, revolving round it. Nevertheless, the observed paths of the sun, moon, and planets, as seen in the sky, were still ultimately explained in terms of complicated circular motions. In fact, forty-eight epicycles were needed—even more than Ptolemy!

This system of Copernicus did have a greater geometrical simplicity, though it still left much to be desired by way of a description that could be seriously regarded as a picture of things as they really were, as opposed to a matter of mere mathematical convenience.

The thought of Copernicus was really only beginning to emerge from mediaeval scholasticism, with its insistence on the Aristotelian concepts of potentiality and actuality, forms and essences, influences, sympathies, correspondences, etc., all related by analogical, metaphorical, and deductive arguments. There was little appreciation by mediaeval science of quantitative measurements such as speed, weight, duration, length, etc., although the importance of observation, measurement, and mathematics had been well understood by the Pythagoreans.

Considerable progress was made by Kepler (1571–1630) after long, arduous, and well-documented studies. He started in a very abstract way: accepting the sun to be at the centre of the solar system, he tried to explain the orbits of the five known planets in terms of various simple ratios. However, nothing quite fitted. He then hit on the idea of some correspondence between the sizes of the orbits and the five perfect 'Pythagorean' or 'Platonic' solids, juggling the latter so that they were appropriately inscribed in the orbits or spheres of the five planets.

This too was unsatisfactory, and it was only Kepler's association with Tycho de Brahe, and accesss to Brahe's magnificent collection of astronomical observations, that led eventually to the establishment of Kepler's three famous laws. These are, briefly, (i) that each planet moves round the sun in an ellipse, with the sun situated at a focus of the ellipse; (ii) that a line drawn from the sun to the moving planet will sweep out equal areas in equal times; and (iii) that the square of the time of revolution of the planet around the sun is proportional to the cube of the average distance from the sun.

It seems probable that, embedded as these laws were in Kepler's voluminous works, he did not himself appreciate their full significance. However, the whole climate of opinion was changing: a change in which Galileo (1564–1642) played a major part. His scientific discoveries, his development of scientific instruments, his enunciation of the laws of motion for falling bodies and projectiles, his implacable hostility towards the Aristotelians in the Universities, and his general taste for polemics and argumentation, all helped to encourage widespread controversy.

The real synthesis of a large number of tantalizing, and often contradictory, fragments came with the work of Newton (1642–1727), whose inverse square law of gravitational attraction explained Kepler's laws in relation to the observed facts of astronomy, and whose Laws of Motion unified both terrestrial and cosmological dynamics. Further advances in the last three centuries have only entailed minor refinements to Newton's work so far as the macro-phenomena are concerned, and the mathematics involved can be handled at secondary-school level.

The above outline could of course be filled in by recourse to the wealth of observational and theoretical detail involved. However, this is unnecessary for present purposes. What is important is that the process of scientific discovery and development can easily follow a long and haphazard course. And although certain clear intellectual principles are involved, uniform progress is by no means guaranteed. Mistakes can occur, and even retrogressions. Moreover, the whole enterprise of scientific discovery is tremendously influenced by personalities, purely contingent events, social forces and technological capabilities.

Nowadays, when the scientific method is used more deliberately and self-consciously, we can specify a number of clearly identifiable and related stages. First, there comes what we may call the 'natural history' stage of fairly uncritical observation. A collection of heterogeneous material is amassed, some of it based on exact measurements, some of it merely anecdotal, and most of it depending on the chance interests of one or more observers. An attempt is then made to classify the available facts and perhaps to construct some systematic description of the whole body of data.

Typical of this stage is the work of a biologist in collecting material on some species of animal. The general appearance, habits and behaviour are recorded, with perhaps special attention paid to methods of obtaining food and the processes of courtship, reproduction, and care of the young. Relationships are noted with special features of the environment, including climatic and geographical aspects, and so on.

Sooner or later, however, some specific questions and problems will arise. Suppose, for example, that two distinct types of colouring, light and dark, are noticed. One may then ask whether these are different subspecies having definite associations with certain kinds of environment, or whether they are essentially polymorphic characters representing a balanced arrangement within a single species? At this stage some convenient hypothesis is formulated which explains the facts in a substantial if not complete way, and which, most importantly, is capable of being falsified by data which might exist and be collected. The principle of verification is central. An hypothesis that is incapable of being disproved, at least in principle, is virtually meaningless. It might of course lead on to more fruitful speculations, but it could much more easily give rise to a whole body of abstract and unverifiable theory. The best hypotheses are those which are simple, of wide application if true, and easily falsified by suitably chosen critical experiments or observations if not. In our zoological illustration we might decide that the two observable body colours were simply due to a single gene substitution at a single locus, with perhaps a corollary about dominance and recessiveness. If the animals in question could be captured fairly easily and if they bred fast enough in captivity, like mice or voles for instance, a suitable series of matings could be devised to test the hypothesis. It might appear that the light form was recessive and the dark dominant, and that a supply of the two forms could be maintained indefinitely by breeding from a segregating line of backcrosses. A certain amount of time would be required before we were reasonably sure that we had found a satisfactory explanation and that no important anomalies were involved.

If some alternative mechanism were operating we might get a fairly early warning from the failure of the segregation ratios expected on a simple single locus hypothesis. Thus on the latter theory we expect 1 : 1 ratios from backcrosses and 3 : 1 ratios from intercrosses. If large numbers are bred it may be obvious whether these ratios are approximately realized, or whether there are large departures from them. Less determinate situations are, however, very frequent. The animals may not be available in very large numbers, and considerable sampling fluctuation may be present. Alternative explanations may then lead to ratios that are not readily distinguishable from those originally expected. With small numbers of offspring we may not, for instance, be able to discriminate at all easily between the 3 : 1 ratio observed in single factor intercrosses and the 9 : 7 ratio obtained from intercrosses involving two complementary factors. It now becomes vital to employ a suitable inferential technique to guide one's further actions. In general, the simple original hypothesis will be rejected if the observations are sufficiently discrepant. This is what a scientific research worker normally aims to do. But because of the presence of considerable natural variation certainty cannot be attained. Appreciable discrepancies between theory and observation may occur purely by chance. In the exact sciences this aspect is usually of only minor importance. With the biological sciences, on the other hand, it is often necessary to use the kind of statistical significance test described in Section 4.2. If the observed discrepancies are large enough, the original, or null, hypothesis is abandoned in favour of some alternative.

This leads us to the next stage in the process. If an hypothesis gives rise to significant discrepancies between theory and practice, it can no longer be retained in its original form. Where there is already strong support for an hypothesis, where for example it derives from an existing well-established theory, the research worker will be inclined to look for some modification of the hypothesis rather than to reject it outright. Suppose, for example, that the animal coat colours referred to above seem to have been established by other workers on a great deal of evidence as due to a single gene substitution with simple dominance and recessiveness. We cannot reject this hypothesis out of hand on the basis of departures observed in the new experiments. We might of course be working with a different genetic factor. This possibility could be tested by crosses between our stock and the animals used by the other workers. If it appears that the factors are the same, we might try supposing that, for some reason, our segregation ratios were disturbed by viability differences between different phenotypes or genotypes. Such ideas must now be incorporated in our hypothesis about the type of inheritance involved, and new experiments devised to examine the credibility of this modified hypothesis.

We have now come full circle, though the new situation is more elaborate and interesting than the original one. Experiments have been performed and compared with other people's work. Additional complications have been introduced, which have had to be tested in turn. Further incidental problems may have arisen on the way, such as small differences in segregation ratios between male and female gametogenesis, occasional 'skipped' generations in the inheritance, or environmental effects like variations in the ratios observed with changes in temperature. Gradually a whole body of knowledge is built up including not only the characters we started with, but also the physical effects of the environment, the behaviour of other characters with which there may be genetic linkage, etc.

While a single isolated hypothesis may be completely ruled out by an experimental contradiction, an hypothesis derived from a well-established theory already has a good deal of support even if it fails to explain new observations satisfactorily. The obvious course is to look for some small modification that will enable the new data to be fitted into the existing theory. In this way the main body of knowledge is kept intact, although subject to modification, change, or development, as required. Not only does the theory supply a convenient and manageable way of handling a wide range of interrelated phenomena, but it also suggests new hypotheses that might be tested leading to new knowledge. From a practical point of view this whole process provides us with continually increasing powers of understanding and control.

The above is in essence what is broadly involved in the scientific method. Its philosophy is basically pragmatic and operational. Hypotheses and theories are only of value so long as they work. Knowledge is never absolute or authoritative, and is valid only so long as it accords with accessible facts. In general, progress occurs by means of adaptation and modification of existing ideas, but sometimes a whole theory must be rejected as no longer tenable. Even then the intellectual framework provided by that theory may have served a valuable purpose in that it has promoted precise discussion and stimulated fresh experiment. Not infrequently there is an eventual return to apparently outmoded theories, and some of their partially valid aspects are incorporated in newer and more comprehensive theories. One thinks, for example, of the primitive corpuscular and wave theories of light, which were eventually followed by a more integrated wave mechanics. Again, the simple particulate, atomic ideas of Gregor Mendel in genetics were subsequently overshadowed by the work of the biometricians at the turn of the nineteenth century, until the neo-mendelian synthesis of R. A. Fisher enabled a firm basis to be supplied for a genetic theory underlying both continuous and discrete characters, as well as initiating extensive quantitative studies of the detailed mechanism of evolutionary advances.

2.2 THE SCOPE OF HEALTH APPLICATIONS

So far, we have undertaken an introductory discussion of the basic ideas of the scientific method. We saw how this had worked out historically in relation to making sense of the apparent movements of the sun, moon, and planets in the sky, and we also considered how the approach could be used in the more complex problems of biology. That such applications can in fact be made on a very extensive scale is of course well-known. Biology is still about the natural world, like astronomy, but with all the additional complications introduced by the multifarious variations in living organisms and by vast networks of interacting processes.

When it comes to medicine and health, another level of complexity is reached. Not only is the biology of man involved, together with his physics, chemistry, biochemistry, physiology, psychology, etc., but there are many other major factors arising out of man-made institutions and socially controlled behaviour. These include questions of morals and ethics, social and political activities, economic structure, and so on. The basic laws of such man-made systems do not have the same compelling degree of clarity and permanence as do the laws describing the workings of the natural world.

The problems of thinking and speaking clearly about these matters are accordingly much more difficult. Working hypotheses are harder to formulate and, even when satisfactorily checked against available data, do not inspire the same confidence as scientific work on the phenomena of the purely natural world.

What then is the role of the scientific approach to health problems? First of all, the field can be divided up into several distinct levels which, though interrelated, are to some extent capable of being treated separately. The most fundamental level is that which deals with the basic biology of man, as already mentioned above. The relevant physics, chemistry, biochemistry, physiology, etc. can all be treated in principle in the customary scientific manner. But there is an additional constraint introduced by the difficulties of carrying out any research that involves experimentation on human subjects.

Such work is nowadays becoming increasingly recognized as requiring careful control because of the inevitable hazards to human life and health. A widely accepted code of practice is the Declaration of Helsinki (as adopted by the 18th World Medical Assembly, Helsinki, 1964, and revised by the 29th World Medical Assembly, Tokyo, 1975). This includes a variety of criteria for proposed investigations that shall respect and guarantee human rights, having proper regard for the relevant moral and ethical aspects. Respect for this Declaration may restrict the experimental procedures available to an investigator, and may call for special forms of statistical design and analysis. However, these restrictions are in no sense antagonistic to the main principles of the scientific method.

The necessity for the scientific method to conform to constraints inherent in the subject matter is nothing new. After all, the scientific work in astronomy, already quoted in Section 2.1, was based (like nearly everything in astronomy) entirely on observation with no experimental interference whatever.

The next level deals with those aspects of medicine that more directly affect people as patients, namely, diagnosis, prognosis, and treatment. At the centre of these activities is the time-honoured relationship between the doctor and his patient. This is quintessentially a situation involving tact, judgement, human feelings, and kindly advice, in addition to the available range of purely technical biological and medical knowledge. The role of the scientific method here is to assist the doctor as decision maker, and at the same time help the patient to cooperate as fully as possible by understanding in nontechnical terms the problem involved and the potentialities for treatment and cure.

Judgement and understanding are needed in deciding into how much detail it is advisable to go. Ideally, the patient should be able to discuss medical matters with his doctor in ordinary language, just as he might consult his accountant or garage mechanic. Of course, it is much more difficult to be dispassionate when one is ill, frightened, or in pain. But most people respond to a sympathetic facing of the broad issues, so that the pros and cons can be weighed up and some course of action followed that has a reasonable chance of success.

In dealing with individual patients, doctors commonly have recourse to a more general kind of knowledge about disease phenomena, i.e. the broad patterns of disease in populations. This could be regarded as fitting into the first level of human biology above, if we choose to include ecology and other population-based disciplines. However, the whole science of *epidemiology*, that is, the study of the incidence, prevalence, distribution, and causation of disease, is now such a basic, extensively developed, and humanly oriented subject, that it is conveniently seen as assigned to a special level of its own.

Epidemiology may not be a very exact science, but it does provide the underlying rationale for understanding the population behaviour of disease. An increasingly scientific approach to epidemiology has supplied knowledge of value both to doctors in treating and advising individual patients, and to public health authorities in controlling the social manifestations of disease.

The biggest successes of epidemiology have so far been in the sphere of communicable diseases. A wealth of biological detail plus the use of quantitative methods of analysis have in many cases enabled the dynamics of the spread of a disease to be described in relatively straightforward terms. Quantitative epidemiological models have been developed for tuberculosis, typhoid, measles, tetanus, cholera, malaria, etc. The results obtained have often led to general insight into the mechanisms of spread involved, and into the likely consequences of alternative public health strategies. In diseases such as measles and malaria the models have been shown to fit actual field data in a statistically satisfactory way. But much remains to be done, as is already evident from the highly disturbing world situation presented in Section 1.1, especially in regard to several widespread parasitic diseases.

With chronic and noncommunicable diseases the state of affairs is less satisfactory. A wealth of information exists, but in most cases the causation is due to multiple factors. We are thus liable to be faced with intricate data, best described in multivariate statistical terms. General control of a disease is made difficult unless we can single out a major factor, or at any rate a small number of factors, that can be easily manipulated.

The statistical correlation between smoking and lung cancer is a case in point. A major correlation can be identified, and although the precise chain of causation may still be somewhat obscure, a substantial method of control exists and would be effective if we chose to use it. Gradually, however, the ever-accumulating knowledge in noncommunicable disease epidemiology, relentlessly subjected to rational and scientific investigation, is gradually providing clues to the complex patterns of causation that will ultimately be of assistance to public health decision makers.

We now pass to the next, more general, level of preventive medicine and public health. It is agreed that it is, in general, better to prevent disease than to cure it, though the relative costs have to be taken into account. The task of public health authorities is accordingly both to facilitate the treatment and cure of individuals and to institute general measures that tend to lower the prevalence of disease in the community and to prevent its occurrence. Although specialized medical and epidemiological knowledge is essential to good public health, there is also a large element of administrative and political activity. To provide a community with what it wants in terms of effective health services, having regard to the constraints of imperfect knowledge, shortages of staff, limited supplies of drugs and equipment, general economic limitations, and the rival claims of other high-priority objectives, is a task of considerable difficulty.

A scientific approach to such problems is still feasible, but the applications look very different from those in physics, biology, or epidemiology. We still need to collect good data, think clearly about them, test our hypotheses, build up a body of reliable knowledge, and see whether the implications lead to effective practical control. But the special forms of the scientific approach are now those entailed in operational research, systems engineering, and systems analysis. It is perfectly possible, and indeed all too easy, to handle these subjects in an abstract and academic way. But the scientific attitude we have been advocating will demand that the tree be known by its fruits: the theoretical techniques are to be justified only by demonstrations of operational effectiveness.

We can now move to an even broader level than public health by seeing health as one of several interrelated sectors, such as economics, agriculture, industry, water resources, energy, education, etc. In some cases health may be improved more easily by appropriate action in another field, e.g. the provision of clean water and effective sewage systems. Alternatively, well-intentioned action in a related field may adversely affect health, as with the consequences of man-made lakes for irrigation encouraging the spread of schistosomiasis (see Section 1.1).

This means that from a broader national level it may be necessary to gauge the overall effect of policies that affect several sectors simultaneously. Ways have to be found of predicting, at least approximately, the probable consequences of any given strategy or policy and of evaluating the net effect of good and bad consequences. This again is a subject for applications of the right kind of systems analysis. It entails the development of multisectoral models that can be used to assist decision makers. The scientific applications are very much in their infancy here, and large claims should not be made as yet: in fact, in most broad policy analysis one cannot so far go much further than a general classification of issues and the thoughts surrounding them, together with a realistic analysis of practical constraints. However, this is certainly a promising field for further investigation, and one of major importance to the whole of society. It is all too plainly evident to any common-sense review of most national planning and decision making that highly unexpected and counter-productive consequences are very frequent. The man in the street usually considers that 'the proof of the pudding is in the eating', and this ultimately is what the scientific approach is about.

2.3 THE LIMITATIONS OF SCIENCE

'The unity of all science consists alone in its method, not in its material,' to repeat Karl Pearson's dictum. 'And not,' we might add, 'in its motivation.' The interest and curiosity of the individual research worker are obviously determing factors, but so are the multifarious opportunities and encouragement provided by the society he lives in. The availability of money and apparatus, a stimulating research climate, and the requirements of the community, will do much to decide what problems are or are not to be investigated. All this is outside the scientific method as such. We can certainly regard this method as the chief and most powerful tool of rational enquiry, but it is still only a means to an end. And ends are chosen on nonrational grounds. 'Reason is, and ought only to be, the slave of the passions,' according to the maxim of David Hume. By 'passion' is meant all the forces of desire, imagination, intuition and morality that determine what ends we want to achieve. Reason and rational behaviour are simply the best means of attaining these ends.

For clear thinking about the scientific method it is important to distinguish these nonrational motivations from the rational form of the process itself. This point is specially important when we come to discuss the function of mathematical models later.

We have emphasized above the nonrational factors that may lead us in a general sort of way to embark on scientific investigations. There is also another point at which nonrational elements play an extremely important part. When we come down to the details of any particular research we find that success in the choice of problems and the formulation of hypotheses is very much a matter that depends on intuition and imagination. The detailed checking of a proposed model for internal consistency and the calculation of predictions is largely a question of deductive logic, but the testing of hypotheses and estimation of parameters involves us in inductive logic, both forms of logical activity being essentially rational. However, the winner of a Nobel Prize for science must not only have reasoned accurately and made efficient use of laboratory apparatus: he must also have had the genius to choose the right problems, and formulate them in a tractable and fruitful way.

No general rules can be laid down for the exercise of intuition and imagination, since these are by definition non- or at least extra-rational. But by seeing more clearly the points at which they operate, we are better equipped to know what aspects of our work will benefit from the more efficient use of rational tools such as scientific methodology, mathematical models, or high-speed computers, etc., and what aspects are decided or determined by quite different considerations.

All these ideas obviously apply with special force to the curing of disease and the promotion of health. If we accept, as discussed in Section 1.3, the premise that health is for people, then it follows that we must start from the wishes, beliefs, aspirations, and convictions of people themselves, whether the latter are considered as single individuals or are viewed as families or broader social groups.

General aims and objectives need to be determined. And commonsense logical analysis is required to see whether there are any inherent mutual inconsistencies in these objectives, and generally to clarify thinking over the whole area. After this we can set about mounting rational and scientifically based endeavours that will help to attain the ends decided upon. More detailed indications of overall health objectives have already been presented in Sections 1.2 and 1.3, together with a brief review in Section 1.4 of the kind of practical activities to be undertaken within a host of social, economic, and other constraints. Section 1.5 emphasized the importance of quantitative methods in facilitating effective actions to some practical health problems, and a scientific approach should be regarded as an essential component in the application of these quantitative methods.

It must, however, be constantly borne in mind that the scientific method, and the range of mathematical skills that support it, are all only technical tools that can help to achieve human goals, and that those goals may be good or evil according to human choice.

CHAPTER 3

The Necessity of Mathematics

3.1 LANGUAGE AND MEASUREMENT

A brief visit to the monkey house of a well-stocked zoo is always delightful and frequently instructive. The impact is perhaps greatest after spending several hours listening to the interminable wrangling of a large committee, or even on occasion after attending the sessions of an allegedly scientific conference! The outstanding points of similarity, apart from the purely visual resemblances, are perhaps the wide range of emotions displayed and the paucity of information communicated. For less serious human activities the gaudy chatter of the parrot house might provide a more suitable analogy. In general, however, the observation of primate behaviour is more rewarding. Everything is there: a continually changing mixture of apathy, curiosity, desire, rage, guile, suspicion, etc. A degree of intelligence and the ability to solve elementary problems of everyday existence are also evident. The greatest single activity conspicuous by its absence is language as we understand it.

The development of language is obviously a vital step in the evolution of a deliberate, conscious control of the environment. Speech is of course more or less essential for the rapid transmission of large quantities of information between the different members of an animal group. And when the stage of writing is reached, there is the possibility of permanent records, whose store of information can be continuously revised and increased. The transmission and storage of information are not, however, sufficient. Decisions must be made about what information is desirable, relevant, interesting, manageable, etc. The great leap forward which the gift of speech has conferred on the human race is not so much in the realm of a primitive kind of data processing, but in the ability to control thought itself. We need not argue here about the degree of thinking present in different species of animals, particularly of course the higher apes. The point is that language provides concepts which can be manipulated independently of the objects to which they refer. If the manipulations are to be successful, in the sense of aiding the control of the external world on the one hand and enriching mental life on the other, certain rules of procedure are necessary. Thus, presumably, language and its grammar evolve naturally. In due course the latter is codified by grammarians, but it is important to realize that they do little more than record actual usage. The conscious and deliberate modification of language to serve specific purposes comes at a much later stage.

To begin with, therefore, language not only provides a means of thinking about the external world, but actually supplies the essential ingredients without which thought would hardly exist. As civilization has developed, so various languages have become finer instruments for human thought and communication. The most striking successes have been in the growth, transmission, and generation of emotional states, as demonstrated by the finest achievements of oratory, poetry, and literature. Strangely enough, these very successes have depended on one of the great defects of ordinary language, namely its imprecision. Although logic has been studied by philosophers since at least the time of the Ancient Greeks, its impact on the majority of human affairs has been remarkably slender. Indeed, the great philosophers seem to have been more concerned with constructing vast imaginative pictures of the world than with providing mankind with linguistic tools of greater accuracy.

The rise of linguistic analysis during the present century holds great promise for the improved understanding of the functions of language, and for a sharper, more accurate use of it. Nevertheless, the practical consequences have so far been disappointing. All too often the linguistic analysts have deserved the gibe that they were learning to speak more and more precisely about fewer and fewer concepts. The main trouble seems to be due to a divorce of theory and practice. Most of those who have been trying to refine language have not been sufficiently involved in the daily application of it to an exacting and rigorous practical activity, such as that afforded by scientific laboratory research. The result is that, in spite of a great deal of philosophical literature on the scientific method, most scientific workers acquire their expertise in the manner of craftsmen by a long practical apprenticeship in following the example set by more experienced men. It should be possible to accelerate this training by a more conscious understanding of scientific principles and their deliberate application in specific practical contexts. A similar kind of advance is now widespread in industry, where new, fast methods of training based on specially developed techniques are replacing the old, slow methods of prolonged apprenticeship.

The situation in scientific research today is urgent. New scientific work is being produced at an ever-increasing rate, and it becomes constantly more difficult and frustrating for workers in any one field to try to keep abreast of new developments in their own subject, let alone in relevant overlapping disciplines. In such a situation there is a premium on the ability to devise and carry out experiments which are really germane to important issues, are economical in time, staff, and materials, are clearly interpretable, and provide results capable of unambiguous transmission to other workers. There have been considerable achievements in this respect in the so-called exact sciences over the past three centuries, and all the prerequisites for handling the accelerated growth of knowledge today in these areas already exist. In the 'softer' sciences of biology, medicine, psychology, and sociology, progress has been slower and more uncertain. An obvious reason for this is that the objects of study are far more variable and complex than those of, say, physics and chemistry. This has made it much more difficult for the biological and social sciences to employ the basic techniques that have been so successful in the exact sciences. What, then, is the fundamental reason why the latter have been able to maintain such a continual and spectacular advance?

Consider, for a moment, once again, what is probably the oldest exact science, namely astronomy. The fantastic achievements of Babylonian and Egyptian astronomy are well known, involving not only the prolonged and accurate observation of the heavens, but an ability to discern significant patterns of change on the basis of which a sufficiently precise calendar could be developed. The latter, with its predictions of seasonal change, was of course highly relevant to the planning of agricultural policy. It is interesting to reflect how, even at this primitive stage, the results of a pure exact science were being harnessed to the practical application of what we might now call farm management. Indeed, the two aspects were clearly interdependent and interacting. To achieve such results it was necessary to do far more than merely look at the heavens and see a mass of brighter or fainter lights. The observation of recognizable patterns and the measurement of their positions was essential. Moreover, the manipulation of such measurements for predictive purposes would imply the existence of some highly competent arithmetical techniques. It is no accident that Babylonian and Egyptian mathematics did in fact possess such qualities. Most probably the mathematics, the astronomy, and the agricultural applications all underwent a simultaneous and mutually interacting development. With the subsequent decline of these civilizations, the dynamic unity of theory and practice was lost, although certain fragments of specific knowledge survived. The same high level of exact science was not reached in Europe until the sixteenth century, but this time, no doubt due to the highly favourable intellectual environment of the Renaissance, progress was rapid and continuous (see, in particular, Section 2.1).

The vital lesson to be learned from this is that even in the ancient world precise knowledge could not be obtained and applied without the basic notions of counting and measurement, together with the arithmetic necessary for handling them. Nevertheless, great civilizations could and did exist in which such skills were rudimentary. We have already noted that the vagueness of ordinary language seems to have conferred actual advantages on cultural activities involving the transmission and generation of emotional states. But even in such contexts relatively precise statements can be made only in terms of concepts that are at least implicitly mathematical. No detailed analysis of a painting or an architectural construction can avoid some reference to ideas of shape, pattern, ratio, space, perspective, weight, balance, hue, etc., all of which entail some degree of numerical measurement and may even go as far as elementary solid geometry. In the same way any discussion in detail about music or literature inevitably introduces some mathematical ideas. It may be objected that all the really important aspects of art and literature are non-numerical. This may be true so far as subjective impressions are concerned. But as soon as we embark on any discussion of such matters that involves more than a mere assertion of personal likes and dislikes we must use language with markedly quantitative elements.

How far one can push this idea is beyond the scope of the present discussion. Nevertheless, it is largely true that most statements intended to convey precise information are capable of some mathematical development. While many people will dispute this if their opinions are solicited, the majority in fact behave as though they believed it to be true. One man will criticize a new building as being ugly. As soon as he says what is wrong, it is a matter of unsatisfactory proportions or measurements, or even inefficient functioning. Another recommends a course of action as being the lesser of two evils. This automatically implies some rough way of measuring and comparing the amount of evil resulting from each action. Again, to come nearer to the main topics of the present book, a writer in a medical journal may complain about the irrelevance of statistics to some problem under discussion. He then proceeds to refute the statisticians by adducing his own facts and figures, looking at averages, comparing percentages, and so on.

It is of course perfectly true that really satisfactory quantification is often extremely difficult. The difficulties are least in the exact sciences like physics and chemistry, and greatest in the spheres of art and morals. Biology and medicine are somewhere in between, and it is the chief purpose of this book to explore, discuss, and illustrate what can be done.

3.2 MATHEMATICS AND COMMON SENSE

In the previous section we commented briefly on the development of language and the emergence of quantitative ways of thinking. Historically, it appears to be incontrovertible that numerical approaches have in fact been sources of enormous power. Arguments could, however, be adduced to show that these applications have all been largely at rather sophisticated technical or intellectual levels, e.g. astronomy and the calendar in the ancient world, or nuclear physics and electronic engineering in the modern world, and that the techniques involved only become widely usable by the masses after they have been duly simplified or packaged for easy application.

There is, of course, something to be said for this view. We all use machinery, e.g. cars or television sets, of whose basic quantitative principles we are largely ignorant. On the other hand, emphasizing this aspect tends to ignore the positive contribution that most people could make to the better organization of their daily activities. Nowadays, one constantly confronts sales clerks who cannot, even with a decimalized currency, assess the total cost of three or four items (or even ten!), having a common unit cost, without recourse to a calculating machine. Yet, only a generation ago in England, it was common for sales girls to calculate mentally the cost of so many yards of material when the unit cost was expressed in shillings, pence, and farthings. Where has all this arithmetical skill gone to? Obviously, the potentiality is still there, but unused and undeveloped.

Nevertheless, the need for a moderate amount of numerical skill is evident in a great many everyday activities. The handling of money, for example, including addition, subtraction, multiplication, division, percentages, discounts, etc. This covers ordinary shopping, income-tax returns, insurance policies and claims, children's pocket money, etc. Again, simple home carpentry and other repair or construction work frequently necessitate some basic arithmetical operations, including the estimation of areas or volumes. Cooking is another everyday activity where a deft handling of weights, measures, timing, and temperature scales is of considerable advantage. Of course, some people do these things more or less accurately in an apparently intuitive way, but if the quantitative basis is intrinsically relevant then it has to be taken into account in some way or other.

I believe that all this shows that elementary mathematical skills are widely distributed in the general population, and that mathematics only begins to seem difficult when it is expressed in an unfamiliar language or notation. Most Europeans find Chinese somewhat impenetrable, but obviously the majority of Chinese children manage quite well when appropriately trained. The abstract, logical languages of mathematics are very difficult for many people, and impossible for some. However, learning any foreign language is difficult, and highly dependent on motivation. The concept being put forward here is that most people, given suitable opportunities and encouragement, could appreciate the basic essentials of ordinary, simple mathematics, because it is so close to their everyday commonsense requirements. But to acquire a facility in understanding and handling the abstract formulations of advanced mathematics requires aptitude plus special training and sufficient motivation.

Again, the appreciation of chance and probability is another matter which is often strongly debated. The mathematical theory of probability entails not only difficult questions of how probability is to be defined, but also a great many technical mathematical manipulations, while the most sophisticated treatises on probability involve discussions of a very subtle nature indeed. What, then, is the relationship between these mathematical treatments and the comprehension of the ordinary citizen?

First of all, there is a good deal of straightforward commonsense appreciation of chance and risk. Practically everybody makes decision choices all the time on the basis of intuitive assessments of relative risks. What is the chance that it will rain and spoil my new hat if I don't take an umbrella? Is it now sufficiently safe to cross the road? Do I have sufficient life insurance? What is the chance of serious indigestion if I eat this food that I find specially delicious, and is the risk worth it? And so on. Admittedly, it may be difficult to put precise numerical values to the probabilities involved, and most people have difficulty in compounding several probabilities. But the basic intuitive grasp of uncertainty is there all right.

Moreover, there are surprisingly large numbers of people who successfully place bets on dogs and horses, and who apparently seem to know what they are doing. Large numbers of people fill in football pool coupons and can make quite short entries which cover very large numbers of permutations and combinations. Others excel at card games like poker or contract bridge, which involve assessing and memorizing certain given data, surmising unknown facts, guessing at alternative possibilities, and devising courses of action that will bring the greatest chances of success. There is in everybody's experience a wide range of practical activities of this kind, demonstrating beyond reasonable doubt an almost universal ability to assess and handle chance phenomena in a commonsense way.

Another practical skill, which is less frequently observed although still distributed through all levels of the community, is the facility of being able to handle large numbers of interrelated processes in an efficient way. This is basically a managerial skill, and it can be observed in operation in the preparation and cooking of a complicated dinner menu, the general running of a family, the management of a business, the control of an industrial concern, and in the administrative and political activities of government. It is, however, a commonplace observation that the more complex of these processes can be very difficult to handle successfully. Those in authority often make decisions that turn out to have effects quite different from those that are intended, making things worse rather than better—the results are in fact 'counter-productive'.

The handling of many complex systems thus works well enough, and often enough, to convince one that there is a fundamental human aptitude at work. But mistakes occur quite frequently when the systems are very large, and the interactions between the component parts are obscure. Reasonable accuracy is, of course, required in forecasting the quantitative consequences of interfering with the system. Here, modern methods of systems analysis and modelling may be helpful, but many of these are still highly controversial and are under current review and development. The essential point to be made is that there is a direct link between the intuitive, commonsense approaches to such problems and the more mathematically oriented techniques which may be suitable in specialized circumstances. This means that the more advanced concepts can be built on commonsense notions, and their implications can be worked out in practical terms as well.

From the foregoing discussion it follows that many ideas and concepts in mathematics are not as remote, abstruse, and removed from everyday life, as many people suppose. There are, in fact, close ties with familiar commonsense understanding, and useful techniques in mathematics, statistics, and systems analysis are capable of being much more widely appreciated and applied than is usually realized.

3.3 LEVELS OF APPLICATION

Several years ago, when the writer was a statistical adviser to the medical director of a small research group, discussions as to the possibility of threading a mathematical path through the thorny intricacies of a multitude of ecological factors were frequently terminated by much dubious headshaking and the assertion that 'medicine is an art'. No one would dispute that this is partially true, in the sense that intuitive skill and imagination are important and essential. At the same time, the majority of patients and potential sufferers undoubtedly hope that the scientific aspects of medicine will continually increase and spread. And science implies mathematics.

The question of the areas of applicability of mathematical methods is a vital one. We have already expressed the view in Section 3.1 that mathematics is involved as soon as we try to conduct any kind of precise discourse, even in such difficult subjects as art and morals. In the current section we shall be a little more specific as regards the range of applications in biology and medicine.

It is well known that one way of looking at the world of nature is to envisage the hierarchy of levels entailed by some such sequence as physics, chemistry, biochemistry, physiology, psychology, sociology. In effect, we start with the basic material elements of the real world at perhaps a subatomic level, and finish with the highly diversified cultural activities of human society. The levels represent a certain line of advance in the direction of increasing organization and complexity. Each level has its own laws of operation so that up to a point it can be studied independently. Nevertheless, the possibilities on any level are circumscribed by what is permitted by activities on lower levels. Thus the laws of physics and chemistry are not suspended for psychology, though the latter involves concepts and laws that go beyond the former. Again, the design and functioning of a health service is on a more abstract level than say human physiology and pathology. The implications of the latter must be presupposed by the former, although the former has a logical content which is partially independent of the latter. We need not labour this philosophical description, or argue points of detail. But the main sequence of levels does roughly correspond to the order of increasing difficulty of applying scientific methods and using mathematical investigations.

As we have already noted, the great and undisputed successes of applied mathematics have been in the areas of physics and chemistry, with which we are not primarily concerned in this book. Mathematical descriptions dealing with biological forms (see Section 3.4) have considerable scope, and are capable of achieving a degree of exactness. Section 3.4 will also open the subject of dynamic models to represent growth, and a start will be made on the problems of handling the fluctuations in size of whole groups of individuals. This entails a great degree of abstraction, but the very simplification involved enables insight to be gained into the laws governing population growth. Statistical variation is also a matter which will have to be faced, though discussion of this will be deferred until Chapter 4.

As we proceed to more abstract levels of the sequence we tend to be confronted not only with an increasing complexity of subject matter, but also with an increasing degree of variation, much of it to a great extent unpredictable. For example, a comprehensive picture of several competing species in a particular environmental setting entails an enormous number of relevant facts. Scientific ecological discussion at a mainly verbal level has made great progress, but real mathematical handling is still at an elementary stage. Again, medical diagnosis often necessitates a physician, in conjunction with other experts, attempting to review a vast array of facts drawn partly from personal experience and partly from the accumulated wealth of medical knowledge in journals and textbooks. The total store of information is increasing at an accelerating rate, and the time has already come with many diseases when it is impossible for a single human mind accurately to survey, value, interpret, and apply all the information available to specific diagnostic problems. Of course, a good diagnostician is one who can make use of long experience and intuitive skill to select a manageable proportion of relevant data and arrive at a sufficiently accurate conclusion. Paradoxically, however, the position is worsening as knowledge advances.

It is in situations like these, where the human mind can barely grapple with the huge complexities and ramifications of the problem in hand even in general verbal terms, that practitioners of the non-exact sciences (including, of course, biology and medicine) often claim that mathematical analysis is inadequate, inappropriate, misleading, or impossible, and therefore best avoided. Although there is a grain of substance in the thought behind this objection, in so far as the existing mathematics may not be adequate, in the long run almost the exact opposite will be true. If a problem introduces a large number of important, interdependent factors, each subject to an appreciable degree of natural variability, it is *only* by using a suitably chosen statistical method that the pattern of correlated measurements can be elucidated, interpreted, and further investigated in a precise way. Similarly, if the number of factors or relevant items of data is so large that they cannot be handled in the human mind, even when some statistical simplification has taken place, then the obvious answer is to appeal to the data-processing capacity of a computer. The utilization of statistical methods and computer techniques is developed further in Chapters 4 and 6, respectively.

The main reason why such mathematical and computing methods are often discounted seems to be as follows. A mathematical model of a biological phenomenon will only seem acceptable to the biologist studying it if he already knows enough about the phenomenon in fairly general, verbal terms to judge the adequacy of the model. Admittedly, this must always be the first, essential, stage. At this point the mathematics is only secondary to the biological thought. It is therefore natural to suppose that as the problems get more difficult and complicated, so the mathematics will fall further behind. What is not always appreciated is that when the mathematics reaches a sufficient degree of sophistication it can move forward under its own laws of thought to supply the biologist with concepts and modes of thinking that were not previously available to him. It is hoped that the truth of this claim will be to some extent demonstrated in later chapters of this book.

So far we have had in mind mainly levels of biological or medical investigation which were distinctly above, though closely dependent on, the more basic levels of physics and chemistry. The subject matter always involved one or more branches of some pure or applied science. Near the top of this sequence would come discussions involving animal behaviour and human psychology. We now pass to aspects which are more sociological in nature, namely the manipulation and organization of the applied sciences to achieve certain more general purposes. This is the realm of what is loosely called 'operational research' and 'systems analysis', which we shall look at in more detail in Chapters 7 and 8. For the moment, we merely note that it is now more a question of applying the scientific method to administrative and executive problems, particularly where decisions must be made about, and interact with, scientific biology or medicine. Silviculture, animal husbandry, general farm management, the design of hospitals, the organization of medical care, and overall health planning, are just a few topics coming into this category.

Of course, not all administrative problems can be handled scientifically using an operational research approach. But many can, and there are great advantages in using such techniques wherever possible, so as to increase the extent of precise discussion and to reduce the area of vague, verbal thought. This enables human insight and judgement to be concentrated on those issues which cannot be settled by more cut-and-dried methods.

Even more difficult than the facilitation of ordinary decision making in human affairs are those matters which are coloured by ethical considerations. Sometimes mathematical analysis can help even here. Difficult problems arise in medicine in the testing of drugs, when a doctor is morally bound to give his patient the best available treatment, but may not really know which is probably best until some experimental trial has been run. Ethical considerations are not suspended, but are in fact made a little easier by a mathematical handling of the technical aspects of the problem.

The main thesis of the present section is that mathematical methods can be applied with advantage to the whole range of activities from basic physics to moral problems. It is convenient, but by no means essential, to think of a hierarchy of levels, or at least a spectrum of activities. As the levels achieve a higher degree of abstraction, so the mathematical approach is harder to apply and is less developed. Nevertheless, properly used, it is not essentially different from the approach of commonsense. It is merely more precise and articulate, has a greater range of usable concepts, and must ultimately be compatible with, though it may go beyond, commonsense verbal reasoning.

3.4 MATHEMATICS IN BIOLOGY

We have already suggested that ordinary language not only enables us to marshal our thought and ideas so that they can be manipulated, recorded and transmitted, but also contributes in a fundamental way to the very existence of the thoughts themselves. This interaction of thought and language, and indeed their mutual development, is paralleled by a corresponding interdependence of thought and mathematics when we come to use the latter for more precise purposes. If we consider something as sophisticated as relativity and quantum theory the truth of this statement is self-evident. Simple illustrative pictures of some of the main ideas and implications are possible, using space–time, travelling clocks, and particles jumping from one atomic orbit to another. Even these require a certain amount of elementary mathematics. But the essential rigorous core of these theories can be handled only in terms of highly elaborate mathematical treatments, without which further constructive thinking would be impossible. Most biological and medical research does not involve any mathematical analysis with this degree of complexity; nevertheless, one has only to consider the application of statistical concepts to the design and analysis of experiments, the use of stochastic processes in studying the growth of biological populations, or the role of information theory in discussing certain aspects of the protein text, to realize that there are many important aspects of biological science in which mathematical thinking does provide dimensions of thought unattainable by purely verbal methods. Some of these applications will be considered in greater detail later. For the moment we merely want to examine more closely the first step of building up mathematical descriptions of biological phenomena.

The great rise of applied mathematics in the seventeenth century had immediate and spectacular successes in applications to the physical world. Consider the range and scope of Newtonian mechanics and optics, for example. Some bolder spirits ventured, even then, to use a mathematical approach in thinking about the forms and development of living things. Thus Galileo discussed the connection between the strengths of materials and the relative proportions of the corresponding structures, not only in the context of building ships or palaces, but also in relation to the construction of plants and animals. He examined, for instance, the limiting height of a tall tree, beyond which a small displacement would result in continuous bending; the effect of increasing the body size of an animal on the dimensions and strength of skeleton and muscles required; and the different physical development that was possible if an animal like the whale was immersed in water with its body-weight counterbalanced by the displaced fluid. Again, in physiological phenomena like the circulation of the blood, a mechanical approach is evident in the work of William Harvey in the seventeenth century, although more precise formulations in terms of hydrodynamical theory had to wait until the nineteenth century. In many such matters it is possible to employ the methods of physics and chemistry together with the appropriate applied mathematics. However, as already explained in the Preface, we are primarily concerned in this book with those aspects of biology and medicine that fall outside the scope of discussions that can be carried on mainly in terms of the exact sciences. At the same time, it must not be forgotten that physical and chemical notions underlie most, if not all, investigations of a more abstract kind such as studies of animal behaviour or evolutionary development.

To begin with, most biological work is largely descriptive. Plants and animals are described with regard to shape, size, colour, behaviour, habitat, similarities to and differences from other organisms, and so on. This is in fact the stage of natural history. At its simplest it is largely verbal and anecdotal, but as observation becomes more accurate and extensive, the possibility arises of using mathematical language to describe or represent the multifarious diversity of living forms. The high point in the development of this kind of descriptive mathematical natural history was marked by the publication in 1917 of D'Arcy Thompson's now classic work *On Growth and Form*. A good deal of this book is concerned with the use of standard statics, dynamics, physics, and chemistry, to provide a more detailed and sophisticated commentary on living organisms than is possible by elementary verbal means. Considerable emphasis is also placed on the competence of suitably chosen mathematical concepts and theories to account for the complexities and intricacies of observed forms.

Consider, for example, the remarkable hexagonal formations in the architecture of the bee's cell, the detailed observation of which dates back at least to Pappus of Alexandria. We can think of the actual configuration as arising from the compression of a double layer of equal cylinders with circular cross-sections. The cross-sections would be expected to assume hexagonal shapes, but more care is needed in examining the connections between the inner ends of the cylinders. It can be shown, as indeed Kepler did in the seventeenth century, that the space-filling symmetry of the whole honeycomb should lead to an arrangement of planes and angles observed in the regular rhombic dodecahedron. Moreover, it turned out later that such a solid has certain optimum properties, namely that the surface area is a minimum subject to certain conditions. From this some have supposed that the bee deliberately chooses this particular form of honeycomb in order to economize in wax. However, further discussion of alternative solid geometrical shapes is a useful corrective to such a speculation. Thus the observed rhombic dodecahedron entails sets of hexagonal prisms with rhomboidal pyramids for bases. Suppose we had flat bases instead. It can be shown that the latter is only about two per cent less efficient. We can hardly suppose that the bee makes such fine calculations, especially when the actual handiwork of the honeycomb turns out on close inspection to be appreciably divergent from the ideal mathematical picture. The walls are thickened, the cells never exactly horizontal, the edges not quite straight, the solid angles blurred with curving surfaces of wax, and so on. It is much more likely that the actual solid geometrical configurations achieved are a joint result of gravity, mutual pressure and surface tension. The important point about such investigations is the enormous power provided by the mathematical description, or *model* as we should normally call it nowadays. Not only, in this particular case, do we have a certain aesthetic satisfaction in finding standard theorems about regular solids so aptly descriptive of a natural phenomenon, but the model can be used as a fruitful basis for further argument and discussion. It is worth noticing that the value and flexibility of such an approach is achieved in spite of the mathematical model being only an approximation to reality. In fact, if the model were made to mimic reality too precisely, we should find the mathematics excessively complicated and unmanageable. More will be said about this conflict of requirements in the sequel.

Another interesting subject which gives considerable scope for mathematical treatment is the question of leaf arrangements in trees and plants, also dealt with in some detail by D'Arcy Thompson. Not only the positioning of leaves on the stems of many plants, but the florets in a sunflower, the woody scales in a fir cone, etc., exhibit remarkable spiral formations which have excited the interest of many biologists and mathematicians. Thus the cones of the Norway spruce have five rows of scales rising sharply up the cone in one direction, and three less inclined rows winding in the opposite direction. With the common larch the numbers are five and eight. Again, the inflorescence of a giant sunflower may show similar patterns in which the numbers are thirty-four and fifty-five, fifty-five and eighty-nine, eighty-nine and one hundred and forty-four, etc. Now, although there are exceptions, the pairs of numbers most commonly occurring are two successive terms in the arithmetical series associated with the name of Fibonacci (or Leonardo of Pisa, *c.* 1170–1250), i.e.

$$1, 1, 2, 3, 5, 8, 13, 21, 34, 55, 89, 144, \ldots,$$

a series characterized by starting with 1,1 and obeying the rule that each term is precisely the sum of the two previous terms. One of the most important mathematical properties of this series is that the sequence of fractions formed by taking the ratios of successive terms, i.e.

$$1, \frac{1}{2}, \frac{2}{3}, \frac{3}{5}, \frac{5}{8}, \frac{8}{13}, \ldots,$$

tends to the so-called *golden section* which is, exactly, $\frac{1}{2}(\sqrt{5}-1) = 0{\cdot}618\ldots$. If a line is divided into two parts by a golden section so that the ratio of the larger part to the whole has this value, then the ratio of the smaller part to the larger is also a golden section. This fraction and its properties were well known to the Ancient Greeks, and it has often played an important part in aesthetic discussions in art and architecture. We may well ask how such a number turns up in the world of nature.

Suppose we envisage the growth of a single plant stem in terms of an ideal circular cylinder, on the surface of which a series of new leaves, A, B, C, D, ..., appear at regular intervals along a spiral path. It may happen that this will lead to an arrangement of leaves lying on two or more vertical lines evenly spaced around the sides of the cylinder. In general, however, this will not be the case. And it is easy to see that it will be natural for the eye to pick out some other spiral such as *A*, *C*, *E*, ... In fact it can be proved from geometrical considerations that there are two complementary basic systems of spirals, one system being right-handed, the other left-handed, and both being distinct from the original genetic spiral *A*, *B*, *C*, ... Moreover, if *O* is a leaf which can be reached from *A* by *m* steps along one spiral and *n* steps along the corresponding complementary spiral, it can be shown that *m* and *n* are successive terms at some point in the Fibonacci series, and that we shall be able to pick out all earlier pairs of terms in the series by suitable choice of spiral paths. While this analysis is approximately true as it stands for nearly cylindrical plant stems or branches, a topologically identical patterning also holds for such structures as the fir cone or sunflower inflorescence by appropriate deformations of the cylinder. It thus appears that the complementary spirals caught by the eye, together with the associated Fibonacci numbers, are an inevitable mathematical consequence of the orderly growth of leaves on the fundamental genetic spiral. In a sense, the appearance of particular arithmetic ratios is a mathematical coincidence. But the unity of mathematics is such that results and theorems following from a given set of assumptions will always be applicable whenever these assumptions can be regarded as true.

Even more space (about a hundred pages) is devoted by D'Arcy Thompson to the equiangular spiral which repeatedly turns up in nature in relation to certain kinds of growth. The simplest possible spiral is one which is traced out by a point *P* as the radius vector *OP* moves with constant angular velocity about the origin *O* while *P* travels outwards along *OP* from *O* with uniform velocity. This is the spiral of Archimedes, and is roughly the form taken by a rope coiled neatly on the deck of a boat, with whorls of the same thickness. In polar coordinates the equation may be written $r = a\theta$. However, in a very large number of coiled shells, such as those of *Nautilus pompilius*, *Turritella duplicata*, *Ammonites*, etc., the successive whorls are by no means equal in size but become progressively larger. The widths of successive whorls are in fact frequently in approximate geometric progression. This is a fundamental property of the so-called equiangular spiral, which may be generated as follows.

Instead of the point *P* moving out from the origin *O* with constant velocity, as in the above description of the simple Archimedean spiral, we now make it move with an increasing velocity proportional to the distance from *O*. The spiral now traced may be represented by the polar equation $r = a^{\theta}$, or $\log r = \theta \log a$, and is called an equiangular, or logarithmic, spiral. It has a variety of interesting properties, including the following. First, the curve cuts all radii at the same angle. Hence one of its names. As already mentioned, the distances between successive whorls along any radius vector are in geometric progression. Also, a succession of radii at equal angles from each other are in geometric progression. Moreover, sectors cut off by successive radii at equal vectorial angles are similar to one another, and may be thought of as arising from a process of enlargement in which a series of continuously growing and completely similar segments are added.

Although the shell itself is not alive, it is formed as the by-product of a growing organism, and one of the simplest ways in which the accretion of new material could occur automatically entails the development of something very close to an equiangular spiral. The *Nautilus* shell is, of course, a three-dimensional structure, but any sequence of corresponding points (e.g. those defining the outline of the shell as it grows) lie more or less in a plane. With turbinate shells, on the other hand, like *Turritella duplicata*, the actual development is in three dimensions. Nevertheless, a similar kind of mathematical description can be adopted, although this time we have to work with the extended notion of an equiangular spiral wrapped on a cone.

In many shells a surprisingly close agreement is obtained between actual measurements and the theoretical values expected from a precise equiangular spiral. Even more remarkable is the fact that in many cases a similar curve will describe fairly exactly the formation of the little lid, or operculum, that closes the entrance to the shell. This lid may grow by successive additions to one part of its edge at the time, while it is simultaneously advancing along the tube of the shell and revolving about its axis so as to provide a tightly fitting plug. A full description of this beautifully adapted arrangement in terms of the appropriate solid geometry is clearly a matter of some subtlety.

In the foregoing illustrations mathematical constructions of a comparatively simple type are seen to describe certain biological phenomena with a surprising degree of verisimilitude. Of course, if one tried to represent the form of a horse instead of a *Nautilus* shell by appropriate geometrical curves, one would quickly come to the view that mathematics was unsuitable for this kind of thing. It is in fact the rich complexity of the world of living things that has led a great many biologists to despair of ever making any extensive use of mathematical methods. However, while it is true that simple and easily manipulated mathematics is of only limited application in biology, this does not mean that elaborate living phenomena are beyond the reach of mathematical analysis. On the contrary, as already suggested in the previous section, we become involved with mathematical concepts as soon as we attempt any kind of precision in the discussion of ideas and the transmission of information. The real question, therefore, is to decide what kinds of mathematical thought are relevant to particular problems, what limitations are entailed, and how such methods may be extended.

It must not, of course, be forgotten that any mathematical description is an idealized, abstract construction, that will at best be only partially realized in practice. Thus we saw that although the cross-section of a bee's cell is more or less hexagonal to a certain degree of accuracy, there are appreciable departures from the ideal geometrical form if we look closely enough. Similarly, a suitably chosen equiangular spiral will broadly represent the shape of a *Nautilus* shell, but careful observation and measurement easily reveal noticeable departures from the theoretical curve. And at best we have only dealt with the main overall form of the shell: an attempted description of the fine structure would present considerably greater problems, not to mention the difficulty of representing the living inhabitant of the shell mathematically.

The more complex the phenomenon under investigation the harder the effort to achieve adequate quantification, partly because some activities are inherently difficult to measure, and partly because excessively complicated mathematical descriptions are too intractable to be of practical value. With a whole organism we are often forced back on such general measurements as weight and size, and a general account of the anatomy and physiology, supported by a more quantitative analysis of certain component parts and modes of functioning. When groups of animals are under discussion even more abstraction takes place. The numbers of animals present may be considered, their ages and sexes, perhaps the range of different phenotypes corresponding to a very small number of genetic loci, certain easily measured patterns of behaviour, and so on. Considerable judgement has to be exercised in deciding what factors are irrelevant, or nearly so, to some particular problem. Quite a lot has to be left out, even in purely verbal biological discussion. Even more is left out in some ways in mathematical biological discussion, but what is included can be handled with vastly greater precision. On the other hand, a complex ecological situation may involve such an enormous amount of biologically relevant information that the human mind is incapable of taking it all into account, even in a general imprecise way, by ordinary verbal processes. It may then happen that statistical methods of compressing and handling the data will be helpful. Alternatively, some of the techniques of data processing now available with large computers may enable fresh progress to be made. These aspects will be examined more fully in later chapters.

It is clear that we are often faced with the conflicting requirements of realism and manageability. A very precise description of all known facts about an evolving species would prevent any useful calculations and predictions being carried through at all. On the other hand, an over-simplified mathematical picture might permit the easy numerical evaluation of certain coefficients, but they would be completely erroneous because some relevant fact had been omitted. This is, of course, one of the perennial dilemmas of all scientific work: it is not confined to biology and medicine, although it does occur here in a specially acute form. More will be said about this important matter in Chapter 5, which deals with the art of modelling.

Dynamic processes

It is a basic characteristic of living things that they undergo birth and death, growth and decay, change and transformation. They are, therefore, deeply involved in dynamic processes of development in time. The world of inanimate objects also exhibits many of these aspects, and some philosophers such as Heracleitus in Ancient Greece have made the notion of perpetual change and flow a central concept in their systems of ideas. Many problems of a purely physical nature, such as the path of planets round the sun, the trajectory of a projectile, the flow of heat or electricity, the oscillatory motion of the tides, etc., can be dealt with in a fairly acceptable mathematical way by the appropriate branch of dynamics. The aspect of change with time usually leads immediately to the formation of differential equations. A wide range of these have now been investigated, mainly in physical contexts, and have provided the applied mathematician with the extremely useful set of tools coming under the general title of 'the functions of mathematical physics'.

Living things, on the other hand, with their self-determination, continuous adaptation, purposive activity, and complicated patterns of behaviour, have proved harder to bring within the ambit of general mathematical laws. Earlier in this section we were concerned with mathematical descriptions of relatively static structures, and looked in detail at the bee's cell, the arrangement of leaves on a plant stem, and the shape of a coiled shell. Even then, we could not avoid some reference to the process of growth and development that had led to the structure in question, and this process naturally coloured to some extent the choice of the appropriate mathematics. But in the present section we want to investigate more specifically situations in which dynamic change and development are completely explicit from the start. Perhaps the simplest of these are the growth of an individual and the growth of a population, matters which were first discussed in a surprisingly comprehensive manner by Quetelet in 1835 in his famous book *Essai de Physique Sociale*.

The use of weight and one or two simple measurements is obviously a rather crude way to chart the development of a single individual. Nevertheless, such methods are widely accepted, and quite rightly, as constituting a useful guide provided that due attention is paid to other factors. Animals and human babies who fail to gain weight at somewhere near the average for the species are suspected as possibly having something wrong with themselves or their diets. This is sometimes the case, but by no means always. Some healthy individuals appear to have their own personal growth curves which are markedly different in certain respects from the majority. The most that can be said is that such a divergence should be treated as an indication that further investigation is required. Again, insurance companies usually view with suspicion any big changes in weight, especially increases, over relatively short periods of time. While there may be perfectly good nonpathological reasons for these changes in individual cases, the evidence from actuarial tables dealing with age, weight, and death-rate is sufficiently well founded to be taken very seriously by any insurance company that wishes to stay in business.

The average healthy baby normally gains weight over the first few months at a fairly steady rate. As time goes by this rate decreases somewhat, although the total weight continues to rise. If observations are maintained over several years, it is found (at least in some communities) that the rate of increase drops to a minimum in boys round about the seventh year, increasing thereafter to a peak in the region of the fourteenth year, and then falling again. The graph is a little more complicated in girls and shows a bimodal curve with peaks at about the tenth and sixteenth years. Similar changes are found in the curves for stature, though they are different in detail from the weight curves, and show an appreciable slowing down just before coming to the main peak, of which there is only one in both boys and girls.

Now all such curves, whether for actual weights and statures, or for their increments, can be accurately described by some mathematical function. Sometimes one finds in the literature that laborious calculations have been performed to fit high-order polynomials with great accuracy. In the opinion of the writer this kind of thing is hardly ever worth the effort. A freehand curve drawn through all the points of the curve probably supplies about as much information as can be usefully handled. In particular, the graphs of increments in weight or stature not only agree with commonsense observations of some periods being relatively quiescent and others involving spurts of growth, especially at puberty, but the figures are closely paralleled by the results of more detailed physiological and biochemical investigations. Thus a single measurement like height or weight does provide some numerical indication of the state of activity of a growing organism, and does in a very elementary way present us with a dynamic picture of the process of development. The difficulty is that crude procedures like this offer little more than mere description. There is hardly enough structure in the mathematical picture to justify the designation 'model', and to enable far-reaching conclusions to be drawn.

Let us now turn to the growth of a whole population. By this we usually mean the *number* of individuals involved, perhaps differentiated by age and sex, rather than the aggregate of any individual physical measurements. It is obvious that the number of organisms in a population is only one aspect of an infinitely rich variety of biological material. Nevertheless, it does provide important clues to the behaviour of the whole group, and is susceptible to a good deal of mathematical investigation. Many biological discussions are concerned with the growth and evolution of species, with competition between species, with the effect of environmental factors, with the spread of disease, and so on. No such discussions can be made at all precise unless they start from some reasonably acceptable mathematical model of the population in question. As always, there is a conflict between having a very simple model that is mathematically easy to handle but is too unrealistic to be convincing, and a very complicated model that has a far greater degree of realism but is too difficult to manipulate and interpret.

One of the simplest models of population growth is that of Thomas Malthus, who observed at the end of the eighteenth century that populations tend to grow in geometric progression. Malthus was concerned because he believed that the means of subsistence could only increase in arithmetic progression, and would sooner or later be inadequate. This is not the place to discuss the economic and social implications of tendencies that were, within limits, correctly observed. The point is that throughout nature most living things do have a propensity for geometrical growth which is only held in check by a sufficient degree of competition, disease, death, and destruction. A typical population grows rapidly at an increasing rate if it starts in an environment with an adequate food supply and a relative absence of predators. As time elapses the food supply becomes less adequate, overcrowding leads to less healthy conditions, fertility declines, and the death-rate increases. Under certain circumstances an equilibrium is achieved, and the population remains more or less constant. It is clear that a knowledge of the precise relationship to be expected between population size at different times and the birth- and death-rates is a matter of considerable interest.

A mathematical form for this typical S-shaped curve of population growth was first devised by Verhulst, a contemporary of Quetelet, using the following kind of approach. First, it is convenient to treat the population size n as a continuous variable, which is accurate enough if n is fairly large. Secondly, we work with continuous time t instead of with discrete generations. Let us suppose that the average rate of growth of the population under favourable conditions is m per individual, so that in time $\mathrm{d}t$ there is an increase of $mn\mathrm{d}t$ in the population size. This means that $\mathrm{d}n = mn\mathrm{d}t$. The behaviour of the population is therefore described by the differential equation

$$\frac{\mathrm{d}n}{\mathrm{d}t} = mn \tag{3.1}$$

with solution

$$n = a\mathrm{e}^{mt} \tag{3.2}$$

if we suppose that initially there are a individuals when $t = 0$. The exponential growth of a continuous population in continuous time, exemplified by (3.2), is equivalent to geometrical growth for discrete numbers over discrete generations.

Verhulst's idea was to impose on the exponential growth of (3.2) some kind of retardation factor which would increase as the population grew. The simplest assumption to make is that the retardation-rate per individual is proportional to the population size, i.e. that the net growth-rate is not m but $m - rn$, where r is the retardation constant. The basic differential equation is now

$$\frac{\mathrm{d}n}{\mathrm{d}t} = mn - rn^2 \tag{3.3}$$

with solution

$$n = \frac{m}{r + (m/a - r)e^{-mt}} \tag{3.4}$$

starting, as before, with $n = a$ at $t = 0$. This is Verhulst's *logistic curve*. Equation (3.4) is in fact a neat mathematical description of an S-shaped curve which rises at an increasing rate to start with, like the exponential curve, but which gradually slows down and finally flattens out to approach the horizontal line $n = m/r$ as t becomes very large. The value m/r is the equilibrium value to which the population size tends.

The intrinsic growth-rate m may well, in practice, be compounded of both birth and death effects. If there is a birth-rate β such that there are $\beta n\mathrm{d}t$ new individuals in $\mathrm{d}t$, and a death-rate μ such that $\mu n\mathrm{d}t$ individuals die in the same interval, then $m = \beta - \mu$, and may be positive, negative, or zero; according to whether the population is increasing, decreasing, or remaining stationary. It is also possible to include many other factors such as migration in a similar sort of way.

Extensive use has been made of the logistic curve in describing the growth of animal and human populations. In demography, for example, attempts have been made to fit such curves to present and past data on the population size of various countries, like the United States, in order to use the prolongation of the curve obtained as a prediction of ultimate population size. While such methods have the positive advantage of sharpening discussion owing to the mathematical basis, considerable care needs to be exercised. Environmental and social conditions may change so that the birth, death, retardation rates, etc., may alter. Thus the model may be approximately true for the period of available data, but this is no guarantee that it will continue to hold in the future. Extrapolation is always an uncertain business, but in the present context some of the difficulties can be overcome by a more detailed analysis of the population, taking account of the birth- and death-rates in a range of age and sex classes.

For a wide-ranging discussion of early work on different forms of growth curves, both for individual members of a species as well as for populations, Chapter 3 of D'Arcy Thompson's *On Growth and Form* is well worth consulting and makes fascinating reading.

An important landmark in the dynamic approach to biological phenomena was the publication in 1925 of Lotka's *Elements of Physical Biology*, subsequently reissued under the title of *Elements of Mathematical Biology*. This book is divided into four parts: 'General Principles', 'Kinetics', 'Statics' and 'Dynamics'. A good deal of it is occupied with matters that are mainly physics and chemistry, but practically every chapter illustrates the important role of change, flow, and process in biology. Even the part entitled 'Statics' deals largely with the equilibria achieved by various continuous processes, whether these are physicochemical like the nitrogen and phosphorus cycles, or primarily biological like inter-specific competition. Lotka's book has had a considerable influence on the development of mathematical biology, and is remarkable for the wide range of phenomena included.

Equations (3.1) and (3.3) above are reminiscent of the equations of motion in mathematical treatments of dynamical systems, an impression which is strongly reinforced by perusal of several chapters devoted by Lotka to 'The Fundamental Equations of the Kinetics of Evolving Systems'. In our very elementary discussion of the growth of populations, we made several simplifying assumptions about growth being continuous, both in population size and in time. Also implicit in the treatment is the idea that any kind of biological variability or random statistical fluctuations were either absent or too small to be of importance. This assumption is often nearly true if populations are sufficiently large. Further, even when populations are subject to appreciable fluctuations, we may still be able to use the equations in question by regarding them as relating to *average* values. There are, however, serious dangers to be guarded against. For some simple processes involving birth, death, and migration, it is quite safe to ignore appreciable variability known to be present, and to develop the equations of motion in terms of averages as though these averages were actual observable quantities not affected by statistical variation. On the other hand, in processes such as arise with competition between species, the spread of epidemics, and in general where there are interacting groups, it usually happens that the true average values are *not* those obtained by solving equations derived on the assumption of no statistical fluctuation. We shall therefore defer any discussion of such situations until the next chapter, where we examine the role played by probability and statistics in the construction of mathematical models.

It should perhaps be mentioned that even the logistic model of the present section, which is similar to a simple epidemic with no recovery, is considerably affected by probability considerations. The *deterministic* kind of approach that we have been discussing should therefore be treated with great caution. Many of the results in the older textbooks, such as Lotka's, have a limited validity, being suitable only in situations where all population numbers remain fairly large throughout the whole process.

The practical value of the models of the present section is that they begin to provide numerical insight into the processes under investigation. The mathematics is developed in terms of parameters, like birth-rate, that have biological meaning. The model can therefore be more precisely checked against the reality it is supposed to describe. Appropriate values of the parameters can be calculated from the data in some suitable way, and used as a basis for further discussion.

3.5 MATHEMATICS IN MEDICINE AND PUBLIC HEALTH

We have already seen in Section 2.2 the general scope of the scientific approach to a wide range of subjects in the health field. In general, the scientific method proceeds most effectively when using quantitative, i.e. mathematical, methods. In Section 3.2 it was shown how such mathematical methods arise naturally from ordinary commonsense ways of dealing with everyday problems, and in Section 3.4 we looked more closely at some of the details of using mathematics in scientific investigations, especially in biological sciences.

It is now time to pay more attention to what, specifically, can be done with mathematics in medicine and public health. We shall, however, be more concerned here with the overall role of mathematics and shall not go into technical mathematical details, as these are beyond the scope of the present book. Moreover, we shall leave specific detailed illustrations to later chapters in which we enlarge on the importance of probability and statistics, the use of modelling, the role of computers, and so on.

First of all, we observe that the wealth of biological knowledge, especially in human physiology, biochemistry, biophysics, nutrition, hormone functions, etc., is completely permeated by mathematical methods of thinking, description, and presentation. The technical work of laboratories in carrying out investigations under rigorously specified conditions, and in relation to established standards, requires the most meticulous attention to quantitative detail. Similarly, research into underlying physical and chemical mechanisms may entail not only quantitative precision but also a considerable amount of relatively complicated mathematics in an effort to test hypotheses about complex processes. Of course, one always expects that the ultimate conclusions will be available in a readily assimilable form.

Thus, a great deal of anatomical, physiological, or pharmacological knowledge can be presented in broad qualitative terms. However, when it comes to specific applications, e.g. the diagnosis and treatment of a given patient, then numerical measurements must be used, tempered by an understanding of what variation may be considered normal and what may be taken as pathological. A facility for handling measurements of temperature, body weight, blood pressure, pulse rate, etc., plus, in certain cases, a host of special laboratory test results, is essential. Moreover, the overall pattern of such measurements and their changes with time also have to be considered.

This does not mean that every general practitioner has to be an accomplished mathematician. But it is obviously an advantage, especially to the patient, if doctors feel at home with such quantitative measurements and can carry out simple calculations based on them. The ability to compare rates and percentages, to assess simple probabilities in relation to the expected outcome of treatment or the risks associated with certain habits or occupations, to adjust drug prescriptions according to the needs of individuals, and so on, would seem to be essential both to good medical practice and to maintaining the patient's confidence in his doctor.

Again, there is a need for doctors to be able to absorb the fruits of research from selective reading of a vast literature. Much of this is highly technical and primarily for specialists. But major questions constantly arise in which patients themselves will raise issues based on their reading of popular press articles, the watching of television programmes, or their consciousness of the claims of high-pressure drug advertising. What does the correlation between smoking and lung cancer really mean? What do the figures imply for me, given my particular history, in terms of increased risks? How much truth is there in the alleged danger, for example, from the protracted use of a particular brand of contraceptive pill? Should one's children be vaccinated against smallpox? Is it really safer deliberately to arrange for girls to contract german measles early in life? And so on, and so on. Threading one's way through a mass of technical information is accomplished more surely by using a systematic method of enquiry that is scientific in its approach and numerical in its technique. If one can build on the joint elements of common sense and mathematics indicated in Section 3.2, then it should be possible to develop a closer working collaboration between patients and their medical advisers.

What we have already said above about special laboratory investigations applies *a fortiori* to specialist treatment, particularly when this involves highly sophisticated forms of hospital treatment. Take, for example, the handling of cardiac patients involving the interpretation of ECG records, the continuous monitoring of some patients using visual display apparatus, the development and application of pacemakers, etc. Or, again, consider the elaborate calculation of optimal X-ray dosages in three-dimensional configurations for the purpose of obtaining a maximum concentration of radiation in tumour growths with minimum effects on surrounding healthy tissues. All such advanced physical techniques usually have a highly structured mathematical basis. The mathematics do not necessarily have to be understood in detail by everyone connected with applying the techniques, but will be an essential part of the underlying processes of control, which are designed to achieve optimal results in individual cases.

Such examples could easily be multiplied to cover most hospital specialities. In fact, wherever intricate forms of treatment are undertaken there is liable to be a considerable degree of 'fine-tuning' in the control mechanisms that necessitates careful quantitative planning and surveillance.

One important component of treatment is a knowledge of the best methods to apply to any given patient. There is a familiar dilemma here. It is a doctor's moral duty to give his patient the best treatment available. But what is best is not always known with certainty: some drugs are a good deal better than nothing, even though leaving much to be desired. What should a doctor's attitude be when a new drug appears on the market? Exhaustive laboratory testing, including animal experimentation, though essential, is not enough. At some point, trials on human subjects have to be undertaken.

The rules governing such trials vary from country to country, but the fairly strict principles set out in the Helsinki Declaration (see Section 2.2) are widely accepted. When trials have been carried out, doctors should be able to interpret the results and decide what is ethical for a given patient. This is not always easy. Proper allowances must be made in a controlled clinical trial for differences in age and sex groups, levels of drug administration, nutritional standards and regimes, environmental factors, responses to treatment, medical history, etc. There is also the complication that patients who respond unsatisfactorily may have to be withdrawn from the trial. This introduces possibilities of unforeseen bias in interpreting results.

Generally speaking, one tries to avoid unconscious bias by the use of placebos along with different drug levels and treatments, plus an allocation of patients at random to the basic experimental groups in the trial. This should preferably be done in such a way that neither the patients nor their immediate medical advisors know until after the trial is over which drugs (or placebos) were being used.

It is obvious that any sound clinical trial is liable to be complex in form and will entail a good deal of quantitative analysis. Because of natural variations in response to treatment and the frequent occurrence of relatively small numbers of patients in at least parts of the trial, conclusions are liable to be couched in statistical or probabilistic terms. The mathematical details of the analysis need not normally be considered by a doctor who is primarily concerned about the results, but a proper interpretation may be highly dependent on an elementary knowledge of probability and statistics.

Another area that entails a fair amount of mathematical support is the field of epidemiology, where we are concerned basically with the incidence, prevalence, and causation of disease. The piecing together of an elaborate jigsaw puzzle comprising large numbers of clues drawn from diverse studies can be an immensely complex business. Much of the material will be statistical data involving rates, percentages, correlations, and associations, all subject to various kinds of uncertainty. Mathematics, statistics, and computer applications are liable to be essential to the relevant research investigations, though it is often possible to present general conclusions in simple tabular form.

In dealing with communicable diseases, an elucidation of the underlying mechanisms of spread usually starts from the broad biological picture. This will be largely qualitative, but with a number of quantitative elements, e.g. incubation period, infectious period, rate of infection, chance of missed cases and carrier-states, variations in immunity, effect of immunization, etc. If population-dynamics studies can come up with models that are simple enough to be understood and which provide a reasonably accurate description of observed phenomena, then it may be possible to make predictions about what is likely to happen under various alternative forms of public health control. More will be said about this later.

For the moment it is sufficient to note that basic research investigations may require a lot of intricate mathematics, though the practical implications may be expressible in relatively simple forms. Even so, it is useful for doctors to have some idea of the quantitative aspects of the infectious mechanisms of diseases like measles, german measles, influenza, infectious hepatitis, etc., in order that rough estimates can be made of the risks involved for individual patients, and so that any treatment carried out can be monitored as accurately as possible. Quantitative epidemiological knowledge may also be useful in making a tentative diagnosis. Precise identification is often impracticable, or even impossible, as with many virus infections, and decisions as to treatment have to be taken quickly. Tentative clinical conclusions are often considerably assisted by utilizing observed patterns of disease incidence current in the community in order to judge, for example, whether a particular patient probably has a common mild upper respiratory virus infection, a more serious and virulent kind of influenza, or possibly an incipient attack of poliomyelitis.

Similar considerations also apply to noncommunicable diseases. But here most of the epidemiological knowledge of disease patterns depends on an analysis of statistical associations and correlations. A large number of interrelated factors may be known to predispose individuals to the occurrence of some specific chronic disease, e.g. age, sex, genetics, nutrition, occupation, environmental hazards and pollution, etc. It is intrinsically more difficult to use this kind of complex information for diagnosis, treatment, or prevention, though there are specialized statistical techniques for doing so. An improved understanding may, however, have far-reaching consequences. This is especially true if single factors with large effects can be identified, as with congenital metabolic diseases related to specific biochemical deficiencies, or the close relationship between smoking and lung cancer.

All this leads on quite naturally to the use of both clinical and epidemiological knowledge for broader methods of control and prevention that can be undertaken by public health authorities. In addition, there is the requirement of trying to improve health services generally at the community level. Any such attempts to plan and organize health control requires a much more sophisticated understanding of the processes involved. The authorities may be faced with calculating the relative pay-offs of different types of action, such as mass screening programmes, various levels of immunization and treatment, changes in nutrition, the use of special health education efforts, etc., having regard to the inevitable constraints in staff and financial resources. The details of the relevant work in epidemiology or health economics can involve a considerable amount of mathematical calculation.

Generally speaking, the implications of any particular control strategy can be presented in commonsense terms. The various costs involved can be specified, and the results expected can be expressed in terms of improvements in the overall disease pattern. The ultimate choice of a strategy is best left to the appropriate decision maker who must consider the available alternatives in the light of various non-quantifiable constraints of a moral, religious, administrative, or political nature. However, because the total situation is so complex, it is unwise for the decision maker simply to accept the technical information as it stands. For a proper interpretation of the material it is essential to understand at least the principles of the quantitative methods employed.

This helps not only in immediate evaluation, but also assists the decision maker to communicate more extensively with his technical advisers and, in particular, to submit them to a more searching cross-examination. We shall enlarge on these aspects in Chapter 8.

It will be noticed that the discussion of this section closely parallels that of Section 2.2. The earlier section was, however, primarily concerned with the development of a broad scientific approach to health problems on many levels. In the present section we have concentrated rather on the possibility of applying mathematical methods to work out the detailed consequences of whatever problems were being investigated.

3.6 THE MULTIPLICITY OF MATHEMATICAL TECHNIQUES

As we have seen, the great strength of science lies in its essentially operational and pragmatic approach. Whatever works will do, at least for the time being. In contrast, however, there is the aesthetic quest for logical unity, elegance, and simplicity. This has always been a tremendous spur to fresh effort, both in thought and experimentation. Whether the universe is inherently simple and uniform, or whether these assumptions are merely convenient first approximations for inquiring minds, is a well-known subject for philosophical debate. Whatever the answer, it remains true that the search for general laws and explanations, for the removal of apparent anomalies, for the construction of models that provide integrated descriptions of phenomena in different spheres of activity, for systems of thought and investigation that have universal application and validity, etc., is a vitally important ingredient of the whole scientific process. There is in fact a certain dialectic of thought and action of which one can take deliberate and conscious advantage.

It is often found that subjects which impinge on a number of other distinct disciplines are specially fruitful in stimulating developments over a wide area. Genetics is an obvious case in point, involving not only heredity itself but also physiology, biochemistry, ecology, embryology, psychology, pathology, eugenics, medical jurisprudence, etc., not to mention the vast array of mathematical techniques that have been developed and utilized in the process. Applied mathematics is itself such a subject. The attempt to solve special problems arising in limited circumstances may lead to the elaboration of general mathematical ideas which react in turn not only on the original problems but also on quite different fields of application. Thus the so-called functions of mathematical physics have their uses in quite different biological and sociological contexts: hypergeometric functions and Jacobi polynomials arise in the theory of the spread of epidemic disease; Bessel functions provide insight into the queueing processes involved in waiting for medical care; and gamma functions are used in computing the most probable of a set of alternative diagnoses.

The importance of mathematics in science lies in the provision of a highly precise language and a system of concepts with which to investigate areas of interest. To begin with, we start with simple counting and measuring, and then gradually introduce more refined ideas according to the requirements of the problems that we encounter. We have already seen something of the applications that are possible in biology, medicine, and public health in the previous two sections. It should be clear by now that although, in a sense, mathematics is a single subject it does in fact encompass a very wide range of ideas, concepts, methods, and techniques. These possibilities have already been extensively explored in the exact sciences, especially physics, but the potentialities for biology and medicine are only just beginning to be realized.

The mathematical approach not only facilitates the precise quantitative description of a problem in terms of some appropriate model, but also provides, or is capable of providing, the means of solving the problem. We have already seen in a general way how such models arise in Sections 2.2, 3.4, and 3.5. The complication that arise in specific instances will be covered more fully in Chapter 5. For the moment it is worth emphasizing that the most critical aspect is the formulation of the model itself. It is usually true to say that, providing a problem is stated with sufficient clarity, some kind of solution is always possible. The mathematics may be intractable from a purely algebraic point of view, but scientifically valuable results may well be obtainable from suitable algebraic approximations, numerical calculations for many special cases, or simulation studies, etc. If the problem is badly stated, on the other hand, or the model adopted is insufficiently realistic, any amount of absolutely exact algebraic development may be only misleading. The crux of all good applied mathematics lies in the choice of an appropriate model to start from, and nowhere is this more true than in biology and medicine.

Assuming, therefore, that we have constructed a reasonable and sensible mathematical model to commence a detailed investigation, we may well ask what methods of procedure are available. It may be that the job is best left to a professional mathematician at this stage, and if this course is likely to be followed it would be wise to consult him during the construction of the model as well, so that he feels more fully identified with the background to the problem and is better placed to interpret subsequent results. Frequently, however, mathematically minded biologists and clinicians will wish to proceed at least part of the way on their own, preferably in consultation with a biometrician or biomathematician. For those who are not primarily mathematicians themselves it is worth reiterating that there is a vast number of alternative and complementary methods of solving mathematical problems. However mathematically unpromising or intractable a model seems to be there is almost certain to be some way of handling it.

As an example of the range of techniques available let us briefly mention a situation arising in the epidemic theory. Consider a closed group of homogeneously mixing susceptibles. A single infectious individual is introduced into the group and the disease begins to spread. In the simplest possible set-up we suppose that there is no recovery (or that recovery takes a very long time and can be ignored so far as the initial stages are concerned). It is easy to construct a simple probability model using standard methods. Essentially, we have a set of differential–difference equations to specify the inherent dynamics of the process. Exact mathematical solutions are possible, but turn out to be very complicated. So far as the actual technique of solution goes, one method is to solve the differential equations in succession starting with the simplest. Unfortunately, this involves very heavy algebra before any general form of solution can be discerned. A more satisfactory procedure is to use the Laplace transform method. The transforms of the required probabilities are quite manageable expressions, although there are complications involved in getting back to the actual probabilities themselves. However, certain mathematical devices are available to facilitate the work. An alternative approach is to use a so-called generating function. This enables the whole set of differential–difference equations to be collapsed into a single parabolic partial differential equation of the second order, which can be solved exactly, though with difficulty. It is possible to deduce many properties of the process from this one equation.

Apart from the purely mathematical investigations just mentioned (which work for the simple epidemic with no recovery, but are much more difficult for most realistic models) we can also consider numerical solution of the various equations. This approach becomes increasingly important with more complicated models. Thus we could solve the set of simultaneous differential equations using standard methods of numerical integration, such as the Runge–Kutta procedure. Again, we could obtain numerical solutions of the partial differential equation for the probability generating function using relaxation methods or a Crank–Nicholson type of iteration, etc. Such techniques all require the use of an electronic computer.

It is clear from the above that there are many different ways of tackling the mathematical solution of any given problem. What we can do in any specific instance depends on the mathematical knowledge and skills immediately available to us and on the computing facilities that can be mobilized. In the long run the multiplicity of techniques is a guarantee that persistent effort will be rewarded with at least partial success.

A further consequence of the vast range of mathematical theory on which one can draw is that not only are there many possible ways of solving a given problem, but there may be alternative ways of reformulating the problem in terms of different concepts. This may provide fresh insights which were not previously available. We shall discuss in Section 4.3 a number of different ways of looking at the technique of drawing inferences from data. In particular, we shall see the importance in certain circumstances of the decision approach, which provides a fairly realistic framework within which to extend the investigation of highly complex administrative and executive processes (apart from the more specific applications to statistical theory). Thus a new analysis in terms of different concepts enables fresh progress to be made.

The possibility of transferring ideas from one field of application to another is always an attractive proposition. If skilfully done, it frequently introduces new and fruitful attitudes. At the same time one has to be careful not to force the material under investigation into a mould to which it is unsuited. For example, most of the basic work in the statistical design and analysis of experiments has been done in the context of agricultural science, to which it is eminently adapted. Much biological experimentation, whether in field or laboratory, is not so very different in general atmosphere. But in the clinical testing of therapeutic drugs, on the other hand, there are many additional complications which are not always easy to encompass in the neatly patterned experiments of agricultural field trials. Some of these difficulties are to do with the multiplicity of side-effects and possible long-term consequences of administering drugs, others are connected with the moral and ethical problems arising in experimenting on human subjects and in the need to provide patients with the best available treatment at the earliest possible moment (see the use of sequential methods in Section 4.4.)

Again, there has been much discussion in recent years about the uses of information theory. No electrical engineer, physiologist, or psychologist would doubt the cardinal importance of this subject. Wherever there is a question of transmitting information in the commonsense meaning of the word, there is considerable advantage in conducting a precise analysis in terms of models which involve a mathematical definition of information. So much is generally agreed. What is under debate is the extent to which information theoretic concepts can be usefully applied to quite different fields of study. Thus a cryptographic approach can be adopted in studying the problems of protein structure and synthesis, involving an analysis of DNA coding in information terms. Such problems are mainly concerned with the storage and transfer of information (suitably defined and measured). It is found that there are often differences of an order of magnitude in the information content when passing from one level of organization to another. We can ask whether this means that the transmission of information is inherently noisy in such contexts, and that accurate results are obtained only by having a high degree of redundancy; or whether the structures that supply configurations essential to basic biological functions also entail, as a by-product, permutations of these configurations involving large quantities of irrelevant information; or whether the additional information entailed is somehow concerned with processing the main message rather than with the message itself; and so on. None of this is, of course, a substitute for precise molecular biological investigation. But it does enable large numbers of quantitative results to be combined and used for further speculation in terms that are themselves based on a quantitative system of rules and concepts, instead of being largely verbal.

It is also possible to study problems of degenerative disease, radiation damage, ageing, and mortality in terms of the loss or destruction of information. Thus organisms will live and multiply in spite of some errors in their genetical information. Radiation may cause an increase in the number of errors, with the result that organisms previously near to the lethal limit will no longer be viable. This may be rephrased in information theoretic terms by supposing that radiation causes gene mutation which means that the noise level in the channel is increased. The channel capacity is reduced to a point where some organisms can no longer transmit a sufficiently accurate message.

When the information concept is combined with notions of negative feedback it is possible to develop many interesting models dealing with the control of whole systems of interrelated functions. Physiological applications are some of the most important, and much light is shed on the mechanism of phenomena such as homeostasis. The relatively new science of cybernetics is built on these foundations, but its field of relevance covers any kind of control phenomena, whether these occur in machines, living organisms or social organizations (see, for example, Wiener, 1948; Ashby, 1952).

So far, it can hardly be claimed that the applications of information theory mentioned above have contributed much by way of new results in biology and medicine that have not already been obtained by other means. Nevertheless, the information concept does provide a useful quantitative tool for speculation and discussion, as well as unifying the ideas and problems arising in a large number of different contexts. As a minimum, this is likely to have a fruitful and stimulating effect; at best, it may lead directly to new developments unlikely to arise from other forms of analysis.

All the mathematical models referred to or explicitly discussed so far in this book have had an obvious numerical basis. They have been derived from commonsense ideas of counting and measurement, have in most cases exhibited only a moderate degree of abstraction from reality, and have entailed some kind of description or formulation capable of fairly easy visualization. The latter characteristic is by no means essential. It usually helps in the early stage of the development of a science, but can be more of a hindrance in advanced work. This is certainly true of much of modern physics, though it is doubtful whether the biological sciences have yet progressed very far in that direction. Counting and measurement are a kind of first-order abstraction that provides the initial impulse for mathematical development. A great deal of mathematics, however, is concerned with higher orders of abstraction. The first step away from purely metrical investigations is the study of the relational aspects between entities. Much elementary geometry is of this nature, and, on more sophisticated levels, there are group theory, the theory of sets, and topology.

Rashevsky (1960, 1961, for instance) and his school have done a great deal to further the study of such applications to biology. The mathematics of this kind of work tends to be rather subtle and difficult, and a detailed discussion is beyond the scope of the present book. Nevertheless, it does provide general insights and theorems which are of great biological importance. Although the practical conclusions have to be couched in largely verbal terms, and formally sound as though they had been reached by imprecise, purely linguistic arguments, they have the great merit of being closely dependent on exact mathematical reasoning. Thus Rosen's theorem in the topological approach may be interpreted to prove that in every organism, however simple or complex it may be, there are some parts that cannot be regenerated. This is exemplified by the fact that the cytoplasm of single cells will regenerate if the nucleus is present, but that the nucleus will not reappear once it has been destroyed. Again, it can be proved that if a non re-establishable component of an organism receives some inputs directly from the external environment, then the loss of that component will result in the death of the whole organism.

The true value of these more abstract mathematical studies has yet to be properly assessed. Simple models are easily discussed, interpreted, and verified. But the greater the degree of abstraction, the harder it is to discover what is being done and whether it is worthwhile. If, over a period of time, the relational types of mathematical model can be seen to have fruitful consequences, as seems highly probable, then the efforts entailed will receive their practical justification.

Another area which has assumed increasing importance in recent years is systems analysis, as already mentioned in Sections 2.2 and 3.5. In fact, many writers regard this as a component of the broader field of cybernetics. Whether we adopt this usage or not, there has been a growing emphasis on the need to grapple with large, complicated systems of interrelated parts in medicine and public health. The subject has therefore tended to develop a certain degree of autonomy, and its pragmatic applications can be developed independently of any general concepts and conclusions in cybernetic theory.

The large systems in question tend to entail complex organization, interactions, and feedbacks between the various components, time-delays in various effects taking place, and even changes in structure due to external manipulation of the system.

Of course, one can usually indicate the general operating pattern of a system by means of a diagram built up of boxes connected by arrows, like an industrial or management flow chart, but such a representation is predominantly qualitative. To study the precise behaviour of the system requires elaborate mathematical description in terms of differential or difference equations. When the number of such equations is large, exact mathematical solutions are liable to be out of the question. Even numerical integration may be too time-consuming. A common method of procedure is then to use the simulation approach, already referred to above, in which computerized calculations in some way mimic the real process (see Section 8.3). This can be a very effective technique, but it has many pitfalls if not used judiciously. It is best combined with certain kinds of mathematical analysis designed to help identify the types of answers to be expected, under what conditions the solutions exist, whether they are likely to be stable or unstable, etc.

The above discussion has deliberately mentioned a number of specifically mathematical concepts in order to give an idea of the range of techniques available. If some of these techniques are unfamiliar to the reader, it will not prevent him from following the subsequent discussions. Many of the ideas will be referred to again later in the context of more detailed illustrations. But in any case it is not our intention to become involved in actual mathematical manipulations: the main purpose of this section is to give a glimpse of the world of mathematics that can be drawn on to help solve problems that are formulated in appropriate mathematical terms.

Once a problem can be quantitatively well-defined, there is almost always some mathematical way to reach at least an approximate solution. Because a great deal of mathematics is already available, and professional mathematicians can invent much more if necessary, we have a tremendous source of potential problem-solving power. Of course, if problems are badly formulated or the mathematics is incompetently handled we can be badly misled. This is all the more reason for the nonmathematical consumer of mathematical technology to understand enough of the general ideas to evaluate proposals and check that sensible operations have been carried out.

CHAPTER 4

The Role of Probability and Statistics

4.1 BIOLOGICAL VARIATION AND PROBABILITY

'If your experiment needs statistics, you ought to have done a better experiment,' said Lord Rutherford. While this might have been true of a lot of physics in the past, it is certainly not generally true in biology and medicine today. Even in the so-called exact sciences 'errors' of observation have long been recognized, but most workers have assumed that such errors could be reduced to negligible proportions by averaging over several repeated measurements. Thus in most physics and chemistry it has been possible to dispose of uninteresting variation, and so concentrate on hypotheses and predictions in which all measurements, even when repeated, could be regarded as more or less precise. This is not to deny that in certain cases, such as radioactive disintegration, Brownian motion, or quantum theory, rather variable phenomena have required special treatments involving the notion of probability.

In the biological and medical sciences, on the other hand, variability is very much larger, more fundamental, and of greater scientific significance. We have already devoted some discussion in Section 3.4 to the subject of growth curves in individuals. If repeated measurements were made on an individual at any one time, small variations of the observational type would easily be detected, but would be of no particular interest. If repetitions were made at short time intervals further variations would be found due to gains or losses from food, respiration, excretion, etc. These items would certainly be of biological significance, but might be adjudged unimportant short-term effects compared with the larger changes over intervals of, say, a week or month that could be related to the overall growth process. Going a step further, we might compare the corresponding figures for different individuals, and this would certainly bring to light variation in the population. Everyone recognizes the fact that individual members of any given species may differ widely in weight or body-measurement, and there is usually not much resistance to the idea of thinking about the average figures as representative of the population. Heights and weights are so familiar to most of us that curves of average growth, or tables showing the average weight for individuals of given age, sex and height, are accepted as providing standards of normality by which to judge any individual case.

But even in the rather pedestrian sphere of heights and weights very large fluctuations can occur as a matter of ordinary natural variation. The writer recalls an investigation into the weights of newborn babies over the first ten days of life in hospital to compare the effects of breast feeding with artificial feeding, taking into account factors such as birth weight, parity, maternal age, and so on. The average curve for a few hundred normal bottle-fed babies climbed steadily over the whole period. The average curve for the breast-fed babies dropped sharply for the first day or two, as expected, then rose rapidly until after a few days it was indistinguishable from the bottle-fed curve. A nice demonstration of nature's ability to overcome the initial shortage of food and achieve a steady rate of growth, one might say. However, the remarkable fact was that although one could attempt to draw general conclusions from the curves of averages, the individual records of the babies appeared completely chaotic: some gained steadily, some lost steadily, others rose and fell, or fluctuated wildly. And there was no obvious relation between these different patterns and the various factors under study. Order and regularity could be clearly distinguished only in averages taken over a large number of individuals. In such circumstances great care would be needed if the overall average curve were to be used as a standard criterion of normal growth against which to judge an individual baby's progress. It would be absolutely essential to take the likely variation into account, so that our basic mathematical model would have to specify not only the average value to be expected at a given age for a given dietary regime, but also some measurement of the variation present.

The use of statistical methods to draw inferences from data involving information about variability will be discussed in Section 4.3. For the moment we are merely concerned with the necessity of *measuring* variability. Reference has already been made in Section 3.3 to the common belief that many biological phenomena are so complex and diverse that mathematical analysis inevitably results in misleading oversimplification. But the truth is that unless a serious attempt is made to develop suitable mathematical treatments, one is merely contracting out of the possibility of a precise discussion.

As is well known, one highly fruitful way of handling essential basic variability is to employ an appropriate probability distribution that specifies the chance that an individual selected at random will be found to have any given measurement or range of measurements. Thus for continuous measurements, such as height, weight, etc., the normal (or Gaussian) probability curve

$$y = \frac{1}{\sigma\sqrt{(2\pi)}} \exp\left(-\frac{(x-\mu)^2}{2\sigma^2}\right) \quad (-\infty < x < \infty) \tag{4.1}$$

where μ is the mean and σ the standard deviation, is often a sufficiently accurate representation (in spite of the fact that, theoretically, it goes off to infinity in both directions). If the curve were used to describe the distribution of stature, then the chance of observing an individual in the range of heights from x_1 to x_2 would be

$$\int_{x_1}^{x_2} y\mathrm{d}x.$$

It turns out that the normal curve is one of the easiest to handle mathematically. Moreover, there are many theoretical reasons for expecting it to be approximately realized in practice, and it is an empirical fact that this expectation is often justified. These considerations are quite sufficient to give the normal curve the position of importance that it occupies in probability and statistics.

With discrete variables, where there is a limited number of alternative observations that may occur, such as the number of albinotic children in a family of a given size, the binomial distribution may be suitable. If there are n individuals, and the chance of any one having a particular characteristic is p (independently of the others), the chance of observing r individuals with the character is the bionomial probability

$$\frac{n!}{r!(n-r)!}p^r(1-p)^{n-r} \quad (0 \leq r \leq n). \tag{4.2}$$

Again, suppose we count the number of radioactive emissions occurring in a given time interval for a specific large lump of radioactive matter, the number of accidents occurring in a given interval under specified conditions, or the number of cells observed in a single haemocytometer square, the most suitable description is likely to be the Poisson distribution for which the probability of observing r events is

$$\mathrm{e}^{-m}m^r/r! \quad (0 \leq r < \infty) \tag{4.3}$$

where m is the average of the distribution.

These are only a few of the commonest and most easily manipulated distributions met with in practice, but it is surprising how much natural variation in biology and medicine can be handled without recourse to more elaborate descriptions. Standard textbooks of probability (e.g. Feller, 1957) or mathematical statistics (e.g. Kendall and Stuart, 1969 and later) should be consulted to gain some idea of the range and power of the available theory.

The use of probability distributions to describe patterns of biological variation is no new device. Quetelet, who was originally an astronomer and meteorologist, was probably the first to use the normal curve in a biological context, in fact introducing it into his discussion of human stature already referred to. Later, Francis Galton (1889) made considerable use of this curve in his statistical studies of natural inheritance, and it played a fundamental role in the extensive biometrical work of Karl Pearson around the turn of century.

Since then, probability distributions of all kinds have been applied to the whole range of biological phenomena including molecular biology, taxonomy, ecology, genetics, psychology, etc.

To begin with, both historically and logically, the probability distributions are no more than sophisticated versions of the mathematical descriptions of Chapter 3. They enable a wealth of diverse observations to be economically subsumed under a single law, which may be characterized by very few parameters: two for the normal curve, only one for the Poisson, etc. This enables the variable phenomena to be more concisely represented and easily thought about. We have in fact what R. A. Fisher has called the 'reduction of the data'. The numerical information can be precisely recorded, stored, transmitted and argued about. Subsequently, these descriptions can be manipulated to construct what is more usually thought of as a mathematical model, namely a structured and developed analogue of reality that can be subjected to the usual processes of the scientific method. This means that consequences and predictions are calculated from the model, tested against appropriate observations, and modifications made to the model if necessary. The testing of models involves us in statistical procedures, which we shall discuss in the following section.

Of course, it has often happened that biologists have been dissatisfied with particular mathematical descriptions, even when these have contained probability elements, on the grounds of oversimplification. Current probability models of competition between species may well look rather primitive to a professional ecologist. The point is, however, that this approach does offer a technique for grasping more surely at the diversity and complexity of nature. And the technique is capable, using modern mathematical, statistical, and computing methods, of being developed to a point where it can do for biology what mathematical physics has done for physics.

4.2 THE USE OF STATISTICS

We have seen in the last section how a wide range of natural variability can often be subsumed under a single suitably chosen probability distribution. This distribution does not specify what any given measurement or observation will be, but does on the other hand make a precise statement about the pattern of events that might occur. In referring to variability, fluctuation, and distributions, we have tended to use the words 'probability' and 'statistical' more or less interchangeably, with more emphasis on the former. It is now convenient to make a certain distinction, though the usages of different writers vary.

Probability distributions and the associated theory are primarily mathematical concepts, though they are derived from commonsense notions of chance and odds, and conclusions must be interpretable in practical terms. In constructing a valid mathematical picture of some phenomenon with strong random fluctuations, we introduce probability ideas and use probability theory to develop the implications of our model. If the model is to be reasonably satisfactory, at least in some respects, then the implications must be verifiable in practice. That is, the mathematical conclusions must bear a certain degree of closeness to observations that have or could be made. It is the function of statistics to be concerned with the relation of theoretical models to reality and with testing their adequacy. In short, while probability theory is still within the sphere of deductive logic, statistics involves *inter alia* the realm of inductive logic.

We have already referred in Section 4.1 to the way in which probability notions can be used to 'reduce the data' to manageable proportions. This means that certain significant aspects are abstracted from reality and the rest are not so much ignored as left undifferentiated within the mass of factors contributing to the essential variation of whatever probability distributions we are using. Whether this simplification is successful or not depends on whether the mathematical model arrived at can be made to yield useful results, irrespective of the fact that factors known to be important in some contexts are not explicitly introduced. It is, therefore, vital to check at an early stage whether the model does fit the data sufficiently well. Because of the natural variation entailed—the reason for the use of probabilities in the first place—no predictions or consequences of the model will be exactly reproduced in practice. However, we shall be satisfied if the correspondence is sufficiently close in a large enough sample of data. Obviously, the criteria of what degree of correspondence is sufficiently close, and what number of observations is large enough, are quite crucial.

At the heart of the statistical testing of a probability model is the notion of a *statistical significance test*. The exact manner in which tests are carried out depends to some extent on which of a number of alternative models of inference one chooses to adopt. It is not surprising that different, though well-founded, models should often lead to the same practical methods and conclusions. This point is taken up again and elaborated in Section 4.3. For the moment we shall merely consider the most widely used so-called 'frequency theory' approach. According to this approach we must sharply distinguish between 'probability' applied to the results of experiments that have or could be made, and 'probability' applied to our degree of belief in some hypothesis. In fact, we reserve the word 'probability' so far as possible to the former context, which corresponds to the way in which probability models or reality were defined in Section 4.1. 'Probability' is thus conceived in terms of frequencies that are, at least in principle, observable.

The mathematical model as constructed constitutes the *null hypothesis*. This may be something quite simple, such as the assumption that the distribution of children in a family between the two sexes is binomial with $p = 0{\cdot}5$ (i.e. a sex-ratio of unity). Testing this hypothesis in relation to a given family involves examining the available data, which may be that there are five children, all girls. We now evaluate the probability of the actual event which has occurred and any other equally or more extreme event that might have occurred. The slightly odd form of this procedure arises from the fact that any particular experimental result may by itself have a very small probability, especially if the number of alternative observable measurements is very large. (With continuous variables the probability of any given reading is, strictly speaking, zero.) We are thus more interested in whether the observed result belongs to a class of uncommon events removed from the most likely result. In the example given above the most likely distribution of the sexes is to find two or three girls in the group of five; one or four being less likely; and none or five being least likely. The observed value of five belongs to the latter class of 'none of five', which has a probability of occurrence of one-sixteenth, or 6·25 per cent, often referred to as the P value.

Now if, in general, P is large, say 30 per cent or more, what we have observed is a fairly common *type* of event. But if it is sufficiently small, say less than 5 per cent, then the observed event belongs to a class of rather uncommon occurrences. We may then prefer to reject the null hypothesis altogether, rather than retain it and be obliged to assume that an improbable kind of event has occurred. The conclusions arrived at by such a technique depend very much on where we draw the line between P values that are acceptable and those that are not, i.e. on the *significance level*. It must be realized, of course, that the use of this type of significance test does not completely insure us against making a mistake. Even when the null hypothesis is true the test will reject it on a fraction P of occasions. If P is too large we shall reject a true hypothesis inconveniently often and shall be inclined to follow up false leads; if P is too small we shall tend to deny ourselves opportunities of rejecting a false hypothesis and developing new ideas. The significance level of 5 per cent is widely employed, and seems on empirical grounds to provide a useful criterion for general purposes. In special cases requirements may be different. Clearly, a more cautious investigator (perhaps testing a dangerous drug) may want to set the level lower, at perhaps 1 per cent or less.

In the example, we see that the observed $P = 6{\cdot}25$ per cent is just not significant at the 5 per cent level. If we are working with this level, we shall assume that the original hypothesis that $p = 0{\cdot}5$ is still acceptable, and that the chance of the next child in the family being a boy is still one-half. If, however, six girls had been observed we should find $P = 3$ per cent, approximately, and should accordingly reject the null hypothesis of a sex-ratio of unity.

The foregoing discussion of significance testing is introduced only to give a simple concrete example, for purposes of illustration, of the kind of reasoning involved. For a more precise account of the theory of significance testing, and the wide range of tests available for application to different situations, the reader should consult any good standard textbook of mathematical statistics. The main point here is that there are numerical techniques for deciding whether a particular mathematical model fits observed data in an adequate manner. If the model does not fit, it must either be modified in some way, perhaps by changing the values of the parameters it contains, or, more seriously, be totally rejected in favour of some alternative description of the phenomenon in question.

So far as the family of six girls is concerned, we might well want to reject the basic hypothesis of $p = 0{\cdot}5$, though we should probably want to retain the implicit assumption that, whatever the value of p, the observed distribution would be binomial. What value, or values, of p are then acceptable? This leads to the problem of *estimation*, another main province of statistics. There are several ways of approaching this problem. One is to find some range of values of the parameter within which there is a given chance that the true value lies. In frequency theory this rule has to be interpreted to mean that if in repeated experiments we calculate 95 per cent ranges, then nineteen times out of twenty the range calculated will straddle the true value. (Notice that on frequency theory only the range can have a probability distribution. The true value of the parameter is some *fixed* though unknown constant.) With a family of six girls, it can be shown that the approximate 95 per cent *confidence range* is 0·61 to 1·00. This range does not include the original null-hypothetical value 0·5, corresponding to the data being statistically significant for that value at the 5 per cent level.

Of course, in this example the sample is very small and the confidence interval correspondingly large. But when samples are fairly large we expect to find much shorter ranges, and these are likely to be more useful for incorporation in other quantitative discussions. Now, as is well known, it often happens that average values have distributions close to normal, even when the individual measurements are not so distributed. This means that we can then use the standard deviation of the normal distribution of the mean as an indication of its reliability. The term *standard error* is usually preferred to indicate the standard deviation of an average value, or other quantity (such as a maximum-likelihood estimate of an unknown parameter) that conveniently has an approximately normal distribution. Thus, if we calculate the recovery rate from a particular disease using a given drug as 25 per cent, with a standard error of 2 per cent, we conclude that in repeated experimental trials of the same size we expect the true rate to lie in a range given by the mean plus or minus two standard errors, i.e. 25 ± 4 per cent, 95 per cent of the time. Speaking more loosely, we feel 95 per cent sure that the true recovery rate is somewhere between 21 and 29 per cent.

The big question about all this kind of argumentation, which is quite familiar, at least at an elementary level, to most research workers in the biological and medical sciences, is how useful are statistical methods in practice and how close are they to the kind of thing a practical scientist is really trying to do? Something has already been said about the logic of scientific reasoning in Chapter 2. Here, we merely observe that most, if not all, scientific research in fact proceeds by hypotheses being set up on the basis of existing knowledge or imaginative guesses and then being tested against suitable experimental data. The hypothesis is retained and developed further if sufficient agreement is reached, otherwise it is rejected or modified. The turning point in the development and application of this statistical approach was probably the publication in 1925 of the first edition of R. A. Fisher's *Statistical Methods for Research Workers*, which has since gone through many further editions. What the Fisherian type of significance test does is to put accepted commonsense practice into quantitative terms. In a very wide range of applications this sharpening of concepts has been found to be exceedingly beneficial. And if in special cases the statistical theory of testing hypotheses has certain difficulties, it at least provides a good basis for the precise development of more adequate methods.

4.3 TYPES OF INFERENCE

In discussing one of the most widely accepted methods of testing for statistical significance in Section 4.2, reference was made to different models of inference. Although in practice the conclusions arrived at rarely differ in an important way with changes of theoretical approach, any general account of the process of scientific research should contain some mention of the variety of logical arguments. We are concerned here not so much with the details of deductive reasoning, in which it is a matter of using language or mathematics to draw logically correct conclusions from well-defined theoretical assumptions, but with the vaguer type of inductive or uncertain inference, in which one attempts to make valid generalizations about the real world.

Most authors of scientific papers today are aware of the simpler types of statistical reasoning, especially as the editors of journals frequently ask for 'the statistics to be put in', presumably to give an aura of respectability which they think would otherwise be missing. It is, of course, liable to be unsatisfactory to try to graft a little statistical thought on to an already completed experimental program. For best results the experimental work should be planned in advance with a view to incorporating suitable statistical techniques right from the start. Nevertheless, an enormous amount of good biological and medical research has been successfully carried out without special attention to statistical subtleties. There are many situations when good statistical design will considerably improve the efficiency of an experiment, giving more information about more factors with fewer observations. On the other hand, poor statistical design implies inefficient, uneconomic experimentation, and may well give rise to misleading results. In such cases new hypotheses, based on these unsound conclusions, will fail to stand the test of time when they are subjected to practical verification. Provided that a constant flow of relevant experimentation is undertaken, most errors that are committed will be eliminated sooner or later by the inherent self-correcting mechanism of the scientific method.

This could explain to some extent the occurrence of fashions in medical treatment. The most charitable view would be that certain new drugs and procedures are seized upon and generally applied because of apparently favourable results due entirely to chance fluctuation in small bodies of data. As the total experience of the medical profession gradually increases with large-scale applications, it is eventually realized that the hoped-for results are not forthcoming. But this process is liable to be slow, uncertain, and uneconomic in operation, and could be largely avoided by well-designed therapeutic trials at an early stage.

It is commonplace nowadays for biomathematical experts to recommend the use of statistical methods in testing hypotheses, estimating parameters, designing experiments, planning surveys, making decisions, or studying the operation of whole complex systems. All of these activities entail constant use of uncertain inference, and it is advisable to consider the status of the alternative methods available. Otherwise, valuable techniques may be neglected or ignored, simply because they are misunderstood or thought to be in conflict with other already accepted procedures.

For a detailed mathematical discussion of these matters some theoretical textbook such as Kendall and Stuart (1969 and later) should be consulted. In the present section we shall merely look at the broad features of the alternative methods. Consider, for example, a simple statistical test of the type described in Section 4.2. Essentially what happens is that we construct a suitable null hypothesis, and then see whether the available data imply the occurrence of a rather rare type of event if the null hypothesis were true. If this does happen we prefer to reject the null hypothesis (at the chosen level of significance), and must then consider some new modified hypothesis which is tested in its turn. We proceed in fact by setting up a series of Aunt Sallies, to be knocked down if the evidence demands it.

Many would claim that most scientists work in this fashion, or at least something very like it, when it is possible to condense the form of complicated judgements into simple terms. But there is an alternative method of testing hypotheses, well known to the users of industrial rejection and acceptance procedures. Suppose that there is a simple alternative to the null hypothesis, the latter being the proposition that some parameter $\theta = 0$, and the former that θ has some other value θ_1. The parameter θ might, for example, be the mean of a normal distribution. Now the Fisherian type of significance test already discussed sets some significance level, say p_1, such that the chance of rejecting a true null hypothesis is precisely p_1, but nothing specific is said about the alternative hypothesis. If we know the alternative hypothesis exactly we can construct a test which entails a significance level p_1, and at the same time allows us to minimize the chance p_2 of rejecting the alternative when it is true (i.e. accepting the null hypothesis when it is false). The quantities p_1 and p_2 are usually called 'errors of the first and second kind', respectively. This approach, in essence, is the basis of the Neyman–Pearson theory of statistical tests. If we can assign reasonable and useful values to p_1 and p_2, the theory provides a very satisfying explanation of statistical tests and a convenient way of deriving new tests when they are required.

In the industrial context of testing samples of some product it is usually possible to make monetary estimates of the losses incurred by committing errors of the first or second kinds. The application of the profit motive then leads to an optimum choice of p_1 and p_2. When there are several alternative hypotheses, or even a continuous spectrum of alternatives, the situation is more complicated, though the basic ideas are the same. The problem in a purely scientific context is to decide whether there is any basis for selecting particular alternative hypotheses. Many people argue that research is essentially open-ended, and that if one hypothesis is shown to be inadequate we must be free to consider possible alternatives in the light of the latest information available, and not to decide in advance the feasibility of any particular alternative. If this view is adopted then the Fisherian approach is a natural choice.

On the whole, scientists do seem to think more in terms of testing specific null hypotheses, and the Neyman–Pearson approach then seems an awkward and unwarranted complication. But since the Neyman–Pearson theory defines its problems more exactly, it is capable of greater mathematical extension and development. Many statisticians therefore maintain that scientists ought to change their thinking so as to be able to take advantage of this more sophisticated approach. Sometimes, in special circumstances, the Neyman–Pearson theory can be applied quite naturally. But in general it is difficult to know what practical implications should be attached to errors of the second kind, even supposing that specific alternative hypotheses can be clearly enunciated. There are difficulties enough in deciding the best probabilities for errors of the first kind, i.e. the ordinary significance levels, though they are easier to interpret in practical terms.

Another approach to all these problems is that afforded by so-called 'decision theory'. Significance tests obviously involve a decision to accept or reject the null hypothesis, and a great many statistical problems of significance and estimation can be couched in decision terms. Whether such methods appeal or not depends to a great extent on the personal predilections of individual statisticians. There are, however, many practical situations, especially those arising in operational research contexts, in which the very nature of the problem involves decision making. Thus a doctor decides to visit one patient rather than another; he opts in favour of one preliminary diagnosis rather than another and is consequently committed to certain further actions; he either treats the patient himself or refers him to a hospital consultant; and so on. If we start asking for a precise account of how to discover the correct or optimum course of action, all kinds of complications arise. We must be able to specify all possible decisions together with the probabilities of the various consequences that might ensue. Further, we shall want some assessment of the value, to the doctor, the patient, or the community, of these consequences. If all this information is available, we may in principle be able to compute the optimum course of action. In practice, of course, human judgement usually has to be relied on to produce rapid decisions on rather inadequate data with the hope that they will achieve fairly good results, if not the best possible.

It might be thought that a decision theory approach to the complex problems of real life would be hopelessly abstract and utopian. There are several reasons for believing that this is not so. Textbooks on decision theory naturally present highly over-simplified versions of real-life problems for the sake of clear illustration, and it may well be that at present such methods can rarely be applied in practice to a problem requiring immediate action, though the further development of high-speed computing facilities may remove this difficulty. Of greater importance, however, is the fact that decision theory provides a framework for the quantitative discussion of highly complicated administrative and executive procedures, which until fairly recently were thought to be beyond the scope of mathematical analysis. It is thus already true to say that the decision approach is providing insight into what happens at such levels of action, and can also help us to be clearer about what we really want. The next stage of trying to discriminate between alternative decision models has as yet hardly begun. So far, therefore, the decision type of model has been mainly of theoretical importance, but it is likely to have increasingly valuable practical applications in the future.

A more fundamental point that arises in connection with the testing of hypotheses is the question of whether hypotheses can be considered to have some kind of probability distribution. Thus in the frequency theory of probability it is only the results of experiments, that are at least in principle repeatable, to which probabilities can rightly be attached. The latter are in fact defined in terms of observable frequencies. Hypotheses, on the other hand, are either true or false, and do not in general have probability distributions, although they are made more or less likely or probable by relevant data. The kind of significance tests suggested by Fisher or Neyman and Pearson involve no assumptions about the existence of prior probabilities of hypotheses. In special cases these prior probabilities may exist. If they do we can use the so-called Bayes' Theorem to determine the posterior probabilities of the hypotheses themselves.

Suppose, for instance, we select for further experimentation a dominant individual from the intercross mating $Aa \times Aa$. The prior probabilities in regard to the new experiments of its being AA and Aa are $\frac{1}{3}$ and $\frac{2}{3}$, respectively, the third alternative aa being ruled out by the dominant phenotype. If we now test-cross to a recessive aa, the occurrence of a single recessive offspring will prove that the individual under test is the heterozygote Aa. But suppose instead we breed three dominant individuals. The likelihood of this happening is 1, or certainty, if the test individual is AA, and $\frac{1}{8}$ if it is Aa. The posterior probabilities are therefore proportional to $\frac{1}{3} \times 1$ and $\frac{2}{3} \times \frac{1}{8}$, or $\frac{1}{3}$ and $\frac{1}{12}$, respectively. Normalizing these relative frequencies to add to unity gives the chances of *AA* and *Aa* as $\frac{4}{5}$ and $\frac{1}{5}$, respectively. We can legitimately say that the odds are 4 to 1 on that the individual tested is a homozygous dominant. Such a statement is really justified only when the prior probabilities exist. In the Fisherian type of test when a significance level of, say, 1 per cent has been achieved, it is often said that the odds are about 100 to 1 against the null hypothesis being true. This is rarely misleading, except to theoreticians, although logically the statement is not strictly accurate.

The inability of frequency theories to supply, in general, odds for or against hypotheses is often felt to be a theoretical disadvantage in contradiction to common sense and common usage. An alternative approach is available in so-called inverse, or subjective, probability theory (see Jeffreys, 1961; Lindley, 1965). This is developed from assumptions that include the use of probabilities to measure degrees of belief and credibility. Thus complete lack of knowledge about a parameter θ, except that it may take any of the values a, b, and c, is represented by equal prior probabilities of $\frac{1}{3}$. An initial preference for the value $\theta = c$ would be indicated by a distribution with a greater concentration of probability at this value, and so on. Although there may be arguments about the correct form of a prior distribution, it does have the advantage of being able to take into account the information already available, whereas it is difficult to do this in frequency-type tests. Subsequent calculations make extensive use of Bayes' Theorem to deduce the posterior probabilities that take into account the information supplied by observed data. The adoption of a so-called Bayesian approach is very popular at the present time.

The inverse probability approach has many attractive features, for example in certain problems of medical diagnosis (see Section 4.7). A typical diagnostic problem is easily and naturally formulated in inverse probability terms, and the final specification of a list of alternative diagnoses with attached posterior probabilities seems closer to actual clinical practice than the methods arising from frequency theories. Of course, as suggested above, we might in such a context prefer to use the decision theory approach, but practical use of this is as yet difficult to implement because too many essential ingredients of the method are not usually available.

We have briefly mentioned inverse probability ideas only in connection with testing hypotheses, but in fact estimation can easily be included. If we consider a continuous parameter θ, then the posterior distribution of θ gives the degrees of belief to be attached to the whole range of possible values. With a fair amount of data this distribution will tend to be concentrated around some preferred value, which can be used as an estimate, the concentration of frequency around this value measuring the appropriate reliability and being represented in many cases by the familiar standard error.

Many people, of course, have a fundamental objection to the use of prior probabilities in cases where they are introduced to measure degrees of belief rather than actual frequencies, as in the genetic example above. In order to overcome this objection R. A. Fisher introduced what he called the *fiducial distribution.* No assumptions are made about a prior distribution for an unknown parameter θ, but the data are used in such a way that mathematical statements can be made about θ which are *equivalent* to a frequency distribution of θ. In frequency theory no such distribution exists, and that is the end of the matter. But in the fiducial approach it is recommended that this fiducial distribution should be used as a measure of the credence to be attached to alternative possible values of θ. The form of a fiducial statement therefore looks just like a posterior distribution in inverse probability theory, though its logical basis is recognizably different. In many simple statistical situations, such as estimating the mean of a normal distribution, fiducial theory, inverse theory, and frequency theory have indistinguishable practical consequences, but differences arise in more complex problems. More serious than this, however, is the existence of several anomalies and apparent logical inconsistencies in fiducial theory. For these reasons many statisticians are highly critical of the whole approach. If the difficulties could be overcome, fiducial theory would be more popular with theoreticians and more readily accepted by practical research workers. The latter are inclined to favour a statistical method which allows them to attach probabilities to unknown hypotheses or alternative values of an unknown parameter, but it is obviously unsafe to use a technique which has logical flaws.

What the outcome of the debate about alternative theoretical approaches is likely to be remains to be seen. In the meantime, how is the practical research worker to decide which of the different highly advertised products to invest in? Will research be more efficient if decision theory is used as a basic informing principle rather than Fisherian significance tests and fiducial theory? Will results obtained by collaboration with a biometrician who favours inverse probability be rejected by a journal whose referees are orthodox frequency men? Such questions ought to worry philosophers and theoretical statisticians. The practical scientist, on the other hand, can avoid interminable debates and the fruitless splitting of logical hairs by adopting the same pragmatic attitude that guides his actions and decisions while working amongst the more familiar technicalities of his chosen field. In this, extensive use is made of models, some largely verbal, others predominantly mathematical (and it is an increasing application of the latter that we are advocating in this book). The alternative theories and techniques of inference can be regarded in precisely the same light, though at a more abstract level. They are simply different models of inference, different models of the way in which valid conclusions can be reached from certain initial data and assumptions. Very often the conclusions turn out to be broadly the same whichever model we adopt, and this is perhaps hardly surprising. Sometimes, however, alternative models, being essentially differently constructed pictures of a complicated activity, lead to diverse conclusions. If these complement one another there is no problem and we have a positive gain; if they are inconsistent further investigation is required, and new insight is likely to be gained by the attempt to resolve contradictions.

The purpose of this section has not, of course, been to give a detailed description of different inferential procedures, but rather to emphasize the multiplicity of methods available, and to draw attention to the possibility that a particular approach may be best suited to a given situation. The approach actually adopted may depend on many factors, including the formulation of the basic problem, the way in which questions are asked, the mathematical theory that is available, the existence of easy access to an electronic computer, and the personal predilections of the investigator himself.

4.4 EXPERIMENTAL DESIGN

The statistical approach of the last section can be adapted to answering all kinds of questions in which we want to know whether some observed average or percentage or correlation is really, or at least significantly, different from a theoretical value. We might simply ask whether an observed recombination fraction between two genetic loci in a series of backcrosses is sufficiently different from one-half for us to be able to assume that the loci are situated on the same chromosome. Alternatively, we might be interested in testing whether the observed figures for male and female gametogenesis appear to differ significantly. Again, we might want to test the recovery-rate from a certain disease against a known standard for untreated cases, or to compare the recovery rates for two different drugs. An outstanding feature of all such statistical analyses is that the whole of the basic, unavoidable, natural variability, except for the specific factor under test, is taken care of by being included in the relevant probability distributions. If this variability is very large we may find it takes a great many observations to yield a firm conclusion, while if it is comparatively small an answer will be obtained more quickly.

Where an effect is influenced by a great many different factors it is quite likely that the basic variability will be large. In such cases it is usually worth trying to distinguish some of these factors, even if they cannot be completely controlled or eliminated. It is often possible to split up the overall variability into separate components, of which one corresponds to the special factor under investigation, several correspond to other identifiable influences, and one corresponds to the remaining undifferentiable variability. Since the latter is inevitably smaller than the original value, a much sharper statistical test is provided. The art of marshalling observations or carrying out specially designed tests to take advantage of these possibilities constitutes the subject of 'experimental design'. For a detailed account of the range of techniques available, the reader should consult some standard textbook such as that of Cochran and Cox (1957). In the present section we shall merely highlight some of the main advantages of a conscious and deliberate designing of experiments.

Suppose, for example, we want to compare the analgesic effect of two different drugs *A* and *B*. We might have 16 suitable patients and decide to divide them randomly (so as to avoid any possible conscious or unconscious bias) into two groups of eight, one group being given drug *A* and the other drug *B*. We then measure the number of hours of relief enjoyed by each patient, and compare the averages for the two groups. If the average for *A* is significantly higher than the average for *B* we should conclude that *A* is the more effective drug. (The actual statistical test used is not important to the discussion. Because of the small number of subjects involved it would probably be some form of 'Student's' *t*-test.) Now, patients vary considerably in their reactions to drugs, and there would be a lot of variability in the hours of relief obtained due to differences between patients. The accuracy of the comparison between the two drugs would suffer accordingly. However, we are not particularly interested in the differences between patients as such in the present experiment, and could eliminate this source of uncertainty as follows. Instead of dividing the patients into two groups, we give each patient drug *A* and drug *B* on two different occasions. The occasions must be sufficiently far apart to avoid interaction, and the drugs should be administered in random order (or possibly in one order for half the patients, and the other order for the other half). For each patient we now calculate the *relative* advantage of drug *A* over drug *B* by finding the number of additional hours of relief obtained. This time we have 16 measurements of relative advantage, and can test if the mean is significantly different from zero. If positive, drug *A* is significantly better than *B*; and conversely if negative. By working with the relative advantage we have eliminated the individual levels of patient response, whether this is high or low, and have in general obtained a more efficient comparison of the two drugs.

This simple 'paired-comparison' test exhibits a very elementary form of experimental design aimed at extracting the maximum amount of information from a given number of observations. Note also that there are other features of the design that require proper attention in practice, but which we cannot discuss in detail here, such as the allocation of treatments at 'random' so as to avoid any unwanted bias, or the spacing of treatments to eliminate interaction effects.

We have seen how just one major identifiable source of variation can be controlled and eliminated in the paired-comparison test. More generally, it is possible to set up *factorial* experiments in which the contributions of a number of factors to the overall variability can be elicited. Some of these factors may be of special interest, while others are merely recognizable nuisances. The philosophy and practice of the new approach, mainly due to R. A. Fisher, were given considerable publicity when his book *The Design of Experiments* first appeared in 1935. Most of the fundamental developments in experimental design were made in the sphere of agriculture. Thus one might wish to compare the average yields of several varieties of wheat, taking account of the use of different fertilizers applied at different levels, as well as variations in soil fertility over areas of ground large enough to lay out the experiment in appropriately sized plots. To begin with, one would try to design an experiment in which all possible combinations of levels of the different factors were included. Thus if we had four varieties of wheat, and three different fertilizers each applied at three different levels, the total number of combinations of levels would be 36. The basic number of plots in a single experimental block should therefore be 36, one for each treatment combination, and it might well be advisable to have at least two complete blocks because of the possibility of variation in soil fertility from block to block.

There are several far-reaching and surprising advantages in adopting a factorial design rather than the classical arrangement of varying only one factor at a time. First of all, one obtains a much more general picture of the effect of each factor, because of the range of conditions provided by variations in the other factors. Secondly, the variety of factor combinations within the experiment facilitates the prediction of results likely to be achieved under a particular combination of circumstances. Thirdly, if the factors included in the experiment have statistically independent effects then as much information is available about each factor as would be obtained if the whole experiment dealt with only one factor, the rest remaining constant. Fourthly, as often happens, the factors are not independent, but have effects that are to a greater or lesser degree correlated. In this situation only a factorial experiment can provide information about the nature of these interactions. When there are known to be several relevant factors, the case for some kind of factorial design is overwhelming. Moreover, there are a large number of special designs available for routine application to constantly recurring types of problem.

Some of the simplest experimental designs consist of several blocks, within each of which all levels of a single factor are tested on the individual plots. This, when properly applied, is a randomized block design. In an agricultural context the blocks might be areas of land in separate fields, and the levels of the single factor might be a graded series of fertilizer concentrations or merely different varieties of wheat. In a laboratory experiment to test the effect of different diets on rats, the diets would be the set of treatments under test and the rats would be the individual experimental units (corresponding to plots of land in the agricultural trial). If we were able to use litter mates and allot one member of each litter to each treatment, the litters would provide the block classification. In the simple paired-comparison test above, a randomized block analysis could have been applied in which the patients were blocks and the drugs were the treatments.

Although it is sometimes difficult to transfer experimental designs developed in one field, notably agriculture, to entirely different fields, the underlying logical pattern is often very similar. It is therefore worth giving careful thought to the possibility that suitable interpretation of the elements of a particular design may enable it to be applied successfully in a quite different context.

This illustrates the power of the mathematical approach to experimental design. Basically, of course, we have to start with some underlying mathematical model. The simplest, and most widely used, model is some variation of the following. Suppose we are trying to examine the effect of just two factors, A and B. Then we assume that the observed effect on some experimental unit of the ith level of A and the jth level of B can be written in the form

$$y_{ij} = m + a_i + b_j + z_{ij} \tag{4.4}$$

where y_{ij} is the observed measurement, m is an overall average, a_i and b_j are the relative contributions of the two factors at the levels specified, and z_{ij} is a random variation imposed on the basic linear, additive scheme. In addition, it is often assumed that all the z_{ij} have the same normal distribution and are independent. Such assumptions are considerable, but as a first approximation frequently justifiable. Thus, if the factor effects are small, only the linear terms will be of appreciable size, and we can ignore possible second-order terms. If the factors do not interact then the form (4.4) is satisfactory. But if there are interactions, further quantities c_{ij} should be inserted. It is, however, possible to carry out a significance test on the basis of (4.4) to see if the interaction terms are required. Again, if the z_{ij} are not normally distributed, it may be possible to work with some other function of the observations, such as square roots or logarithms, that restore approximate normality.

From the elementary form of (4.4) it is easy to develop models involving large numbers of factors, blocks, interactions, and other sophistications of patterning required by practical exigencies. The point is that for a very large number of available designs the actual form of the computations required is relatively simple and straightforward, usually involving the repeated calculation of certain sums and sums of squares of appropriately chosen experimental measurements. This leads to the well-known presentation of results in an 'analysis of variance' table, from which the significance of all the distinguishable aspects of the experiment can be deduced.

One aspect of modern experimental design which may be singled out for special comment is the possibility of a sequential ordering of the experiment. In the standard type of experiment we have to decide in advance how many observations to collect. If these turn out on analysis to be too few we must try to obtain more, which may be difficult or impossible at that stage. If it transpires that too many observations have been made to achieve the desired accuracy then time and money have been wasted. In medical contexts these difficulties are intensified. No doctor wants to prolong an experiment longer than strictly necessary, and wishes to give his patients the best treatment as soon as this is known. Ethical considerations thus become inextricably woven into the choice and design of the experiment. Now a sequential method means that the experimental units, which may be patients, are treated serially. An assessment of the situation is made at each stage in such a way that a decision can be reached immediately on whether to adopt treatment A or treatment B, or alternatively to continue experimenting because no result is yet possible. In this way the experiment can be terminated at the earliest possible moment, and in general is appreciably shorter than would otherwise be required. Moreover, in medical contexts the running of an ordinary experimental trial is often made very difficult or even impossible by agonizing discussions following a few failures, perhaps deaths of patients, about whether the trial should be discontinued or not. The sequential arrangement means that alternative courses of action, given different experimental results, can be considered carefully and calmly in advance. The best decisions during the course of the experiment are then made much more easily, and combine the dictates of morality and statistical efficiency. For a detailed discussion of sequential methods in medicine see Armitage (1960).

There is no space here to develop further this discussion of experimental design, which already commands a huge literature. Our main purpose, however, has been to show how a simple mathematical model of the joint effects of several factors in an experiment, including a possibly large random probability element, can be used to disentangle the quantitative actions of these factors in a precise and highly informative way. Thus the simple statistical tests described in Section 4.2 can now be applied to gauge the significance, and estimate the extent, of complex patterns of highly variable and simultaneously acting factors—precisely what is required for the enormously diverse phenomena investigated by biological and medical science.

4.5 STOCHASTIC PROCESSES

The central importance of dynamic processes of change and development to all biological investigation was emphasized in Section 3.4. We saw how certain salient features of growing populations could be abstracted from a highly complex reality and described by some relatively simple differential equations. In the illustrations given these equations were easily solved to yield exponential and logistic curves. Although considerable simplification was necessary in the construction of the mathematical models, it was claimed that some insight into the biological processes was achieved, and that a basis was provided for further investigation. In particular, it was pointed out that the deterministic models of Section 3.4 had the great defect of ignoring basic biological variability (as distinct from secular changes with time), or at least assuming it to be negligibly small. Where numbers are sufficiently large, deterministic models may be adequate within the limitations imposed, but in many important situations, epidemics and interaction between species for instance, some numbers are liable to be quite small at certain times. In these circumstances statistical fluctuations are liable to exert a special influence on the behaviour of the whole process.

For making demographic predictions of population structure a deterministic analysis based on growth-rates by age and sex classes may be perfectly adequate. Provided the estimates of population size a few decades ahead are accurate to within a few hundred units, they will suffice for most economic or sociological purposes. But suppose, on the other hand, that we are trying to make a quantitative study of ten breeding pairs of some rare species of bird. It is obvious that chance will play a major role in regard to such factors as fertility, clutch size, food supply, climate, competition, disease, accident, and death. Any description of the possible changes in the small population with time must take account of the fact that things could turn out quite differently. This does not mean that a precise treatment in mathematical terms is impossible, but rather that a probability model is essential to describe the patterns of happenings that could occur with their relative chances of occurrence. We must therefore envisage the existence at any point of time of a probability distribution of population size, instead of the single value specified by a deterministic model. We now have a *probability process* or *stochastic process* in contradistinction to the previous *deterministic process*.

Similar considerations would apply to charting the course of an epidemic in a small family group, of say four children. It would be of little interest to know even the average number of cases after a given time in such a small group. But if we could make probability statements about the actual times of occurrence of the different cases, distinguishing between susceptibles, infectives, and those who were isolated or recovered, and taking into account the existence of latent periods and variable infectious periods, then we would be well on the way to having a biologically meaningful description of the whole process. Such a model would not only be an aid to epidemiological insight, but would form a sound basis for making probability predictions about the likelihood of new cases occurring. Moreover, theories of epidemic spread through a whole population could be more securely related to individual clinical and epidemiological observations. For the moment we shall merely look in a general way at what is involved in trying to handle a whole process of probability distributions.

Let us examine briefly the stochastic analogue of the simple case of a process involving only birth. In the deterministic case we should assume a birth-rate β, say, such that a population of size n would receive an increment of $\mathrm{d}n = \beta n \mathrm{d}t$ in time $\mathrm{d}t$, leading to the exponential law

$$n = a\mathrm{e}^{\beta t} \tag{4.5}$$

(assuming $n = a$ when $t = 0$) as already indicated in equation (3.2). Various modifications are possible in order to introduce probability elements into this situation. The simplest is to assume that the chance of a given individual reproducing in time $\mathrm{d}t$ is $\beta \mathrm{d}t$. The chance of one new individual in the whole population in time $\mathrm{d}t$ is therefore $\beta n \mathrm{d}t$. We now work directly with the actual discrete population size, and write $p_n(t)$ for the probability that the population has precisely n individuals in it at time t. The population size at time $t+\mathrm{d}t$ can be related to the size at time t by fairly elementary considerations. For if there are n individuals at time $t+\mathrm{d}t$, there must have been either n individuals at time t and no change in $\mathrm{d}t$, or $n-1$ individuals at time t together with one birth in $\mathrm{d}t$. Combining the probabilities, we can thus write

$$p_n(t+\mathrm{d}t) = p_{n-1}(t).\beta(n-1)\mathrm{d}t + p_n(t).(1-\beta n\mathrm{d}t)$$

which gives, on rearrangement and division by $\mathrm{d}t$, the differential equation

$$\frac{\mathrm{d}p_n}{\mathrm{d}t} = \beta(n-1)p_{n-1} - \beta n p_n. \tag{4.6}$$

This is valid for $n > a$. It is easily seen that the corresponding equation for $n = a$ is slightly simpler, namely

$$\frac{\mathrm{d}p_a}{\mathrm{d}t} = -\beta a p_a \tag{4.7}$$

since there can be no term involving p_{a-1} as the process starts at the value a.

Sets of differential–difference equations like those given by (4.7), which may be regarded as the *equations of motion* for the stochastic process, are often somewhat intractable in general. Several techniques of solution are, however, available. Routine, though rather technical, mathematical manipulations lead to the required solution

$$p_n(t) = \binom{n-1}{a-1} \mathrm{e}^{-n\beta t}(\mathrm{e}^{\beta t}-1)^{n-a} \quad (n \geq a) \tag{4.8}$$

which can be recognized as a special case of the so-called negative binomial distribution. This distribution therefore replaces the single value $n = a\mathrm{e}^{\beta t}$ of the previous deterministic treatment.

The mean and variance of the distribution can be deduced from standard theory. They are, in fact,

$$m(t) = a\mathrm{e}^{\beta t} \tag{4.9}$$

$$\sigma^2(t) = a\mathrm{e}^{\beta t}(\mathrm{e}^{\beta t}-1). \tag{4.10}$$

It is immediately noticeable that the stochastic mean in (4.9) is identical with the deterministic value in (4.5). With large numbers, therefore, the deterministic treatment would be a successful substitute for any stochastic discussion that was primarily centred on the mean values. On the other hand, when numbers are small, e.g. when the initial population size a is at most a few units, the amount of variation present, as represented by the variance (4.10), can be considerable. Stochastic means are often used as a first step towards describing the behaviour of a stochastic process, so that it is clearly convenient when the more easily calculated deterministic value is the same. This state of affairs is, as already mentioned, by no means always the case, and should not be relied upon.

Some care is needed in the interpretation of (4.8). If we chose to observe in detail any particular population for which the model was suitable, we should of course have just one observed figure for each instant of time. A graph of population growth would show appreciable fluctuation, and the question arises of how this fluctuation is related to the stochastic probability distribution. The meaning of (4.8) is that, if we imagine a large number of populations all starting off simultaneously with a individuals, then the proportion of these populations having size n at time t is theoretically $p_n(t)$, with average size $m(t)$ and variance $\sigma^2(t)$. Any single population, or realization of the process, might well show considerable divergence from the stochastic mean curve, although we can use the latter in conjunction with the known variance as some measure of the variability presented by the whole process.

We have looked at some of the mathematical details of a stochastic treatment of the simplest kind of birth process in order to show the sort of approach that is involved in handling and thinking about processes with an appreciable probability element. When extensions are made to include losses from death or other causes, at least one new aspect arises, namely the possibility of extinction. Thus the simplest deterministic treatment of a population with birth and death leads to equation (3.1), where m is the average growth-rate. If the birth-rate is greater than the death-rate, m is positive and the deterministic size increases exponentially. But in a stochastic treatment there is always the possibility that a sufficient number of deaths will occur by chance for the population to become extinct. (Of course, with negative m the deterministic population declines steadily to zero.) This probability of extinction is an important feature of the process, unaccounted for in the deterministic model.

Again, there may be other circumstances in which the properties of stochastic models are quite different from deterministic. The phenomenon of recurrent epidemic outbreaks is quite inadequately treated by a deterministic treatment, which predicts a series of damped oscillations that are not observed in practice, while the stochastic theory does provide for a recurrent series of fluctuating but undamped outbreaks.

It should now be clear from this brief résumé that the theory of stochastic processes is of central importance in the study of biological phenomena. It is also clear that even very simple stochastic models are liable to engender an unexpected amount of mathematical manipulation. The situation becomes increasingly difficult as one attempts to make the models more realistic. In the birth process examined above, for example, we made the very convenient assumption that the chance of an individual reproducing in time $\mathrm{d}t$ was $\beta \mathrm{d}t$. Now it is easily shown that this means that the time interval between birth and reproduction must have the negative exponential distribution $f(t) = \beta e^{-\beta t}$, $t \geq 0$. This distribution represents more variability than is usually found in nature. In bacteria, for example, a unimodal peaked curve approaching a normal distribution is nearer reality. We must therefore be prepared to introduce a good deal of complication into stochastic models if they are to describe actual phenomena sufficiently closely.

There is a dual problem here. Firstly, an oversimplified stochastic model may be very useful in that it gives some indication of the properties to be expected of models with probability elements, as opposed to those without. This happens, for instance, in the theory of epidemics in large populations. Secondly, the statistical testing of models against real data and the estimation of biologically meaningful parameters must be carried out with models that have the necessary degree of realism. With epidemics this can actually be done with small family groups, but in general the more realistic and complicated models lead to highly intractable mathematics. This problem will be partly solved in due course by new mathematical advances, particularly in the theory of approximations. The importance of modern computers in this context cannot be overestimated. On the one hand these can effect the numerical solution of the highly complicated systems of equations describing a realistic stochastic process, and on the other hand they make possible extensive investigations of a *simulated* process. That is, the computer produces a large number of artificial realizations of the process, based on the mathematical model, and then computes the means, variances, probabilities of extinction, etc., to some acceptable degree of accuracy. The role of simulation and allied methods will be considered below.

In this brief sketch of the role of stochastic processes in biology we have confined the examples to processes which involve populations of individual units changing with continuous time. It is often, however, more convenient or suitable to work with discrete time, as in population with non-overlapping generations. Again, in diffusion approximations to discrete-variable processes, both the population size and time are continuous. There is in fact a wide range of different kinds of model to choose from, and some acquaintance with the alternatives is very desirable for anyone working with biological or medical processes. For a relatively simple applied mathematical (and nonrigorous) account of the elements of stochastic processes the book by Bailey (1964) may be consulted. More advanced treatments are available in Bartlett (1966), Bharucha-Reid (1960), Parzen (1962), and Cox and Miller (1965), while several specific applications to biology and medicine are presented in Gurland (1964).

The simulation technique referred to above is essentially a statistical sampling method. If we can handle a stochastic process by an entirely mathematical analysis, then, in calculating any means, variances, or probabilities that we are interested in, full account is automatically taken of the whole range of possible variation. All that a simulation study does is to construct a sufficiently large number of possible realizations of the process in question, and then calculate the required quantities from this limited sample of quasi-observations. These quantities will of course not be obtained with the absolute accuracy of purely mathematical deductions from the model, but, as is usual with statistics calculated from samples, will have a degree of imprecision indicated by an appropriate standard error. Because the results obtained from a simulation study are to some extent blurred by sampling variation there is a certain loss of definition, so to speak, in the perception of underlying patterns of behaviour. Moreover, the simulations have to be performed for fixed values of parameters in the model (e.g. population size, birth and death rates, number of generations observed, etc.). In order to get a reasonable cover of alternative combinations of parameter values, quite large numbers of simulations may have to be used. Even then critical relationships between different parameters in the model may be lost. A purely mathematical investigation, on the other hand, would, provided it were tractable, reveal the basic structure and implications of the whole process in a much more fruitful and comprehensive way. For these reasons it is often said that the use of simulation is a confession of mathematical failure. While this charge may be admitted, it may be riposted that complete and elegant mathematical solutions are very often available only for grossly oversimplified models. The proper use of simulation therefore lies in knowing when it is necessary to supplement more theoretical studies. In fact it may well turn out that the results of a simulation investigation will suggest that fresh theoretical efforts be made in a particular direction, thus providing a synthesis of two contrasting approaches.

Elementary stochastic birth-and-death processes are very useful for investigating the kind of properties to be expected in various ecological contexts when we abandon rigid deterministic models and introduce a range of probability elements. Nevertheless, simplifying assumptions that entail, for instance, the negative exponential distribution of lifetimes or the absence of a latent period in the development of an infectious disease are likely to have only very limited practical applications. They may provide some general insight, but will not apply in detail to any actual situation of interest, and will not in particular permit the estimation of biologically important parameters.

To achieve the latter demands a fair measure of realism in the model. We may be forced to assume that lifetimes have an approximately normal, or perhaps gamma, type of distribution; that an infected individual passes through latent and infectious phases, having some appropriate joint probability distribution; that populations have an age and sex structure with different characteristics for each group; that there is a spatial diversity to be taken into account; and so on. As soon as the least attempt is made to incorporate such factors into the basic mathematical model, explicit mathematical treatments are usually ruled out, or at least cannot be developed very far. It is still of the highest importance to be able to formulate the more complex model in well-defined mathematical terms, as this is the only basis for precise, rational discussion. In some cases, convenient mathematical approximations may be possible, but in general computerized simulations will be the only possible way of advance.

Simulation has already been applied to a wide range of biological and medical phenomena, including the interaction between species, the evolutionary behaviour of genetic systems, the spread and recurrence of epidemic disease, the functioning of neural networks and other physiological systems, etc. Another fruitful area of application has been to the field of operational research, already mentioned briefly in Section 2.2 and to be dealt with more explicitly in Chapter 7. Here we are concerned more with administrative and executive problems. And insofar as such problems of everyday life admit of a realistic mathematical formulation, the latter usually has to be fairly complex and to include all kinds of inelegant and contingent facts. This rarely makes for tractable mathematics, and simulation studies of organizational systems are essential if real progress is to be made. Further aspects of simulation in the context of operational research are discussed in Section 7.3. Some specific examples of the application of operational research methods to medicine are elaborated in Sections 7.5 to 7.7, particularly in relation to the function and design of hospitals and the study of complicated queueing processes in outpatient departments.

The term 'Monte Carlo' is often used synonymously with 'simulation', although in origin it involved a logically rather different idea. The simulations we have been considering are realizations of a stochastic model which are strictly analogous to possible realizations of a real-life process. In contrast to this the Monte Carlo approach (the actual name being a somewhat transparent code word) was developed during the Second World War for solving very difficult mathematical equations, arising for the most part from purely deterministic problems, by probability methods. The technique was to invent a probability problem, whose treatment led to the same equations as the deterministic problem under investigation, and then to solve the probability problem by sampling experiments which were essentially simulations. A very simple example of this is the evaluation of the area enclosed by some plane closed curve. We surround the curve by a square of area A, and use a standard statistical method to choose points at random within the square. If the proportion of points lying within the curve is p, then pA is an estimate of the area contained by the curve. This idea can easily be generalized to multidimensional problems, for which conventional numerical methods are liable to be extremely laborious and to require huge numbers of points to obtain any answer at all. Moreover, these methods only achieve accuracies proportional to $n^{-1/d}$, where d is the number of dimensions involved. The Monte Carlo approach, on the other hand, will give some answers for any value of n and will achieve estimates of accuracy proportional to $n^{-1/2}$, which may be highly expeditious for large n.

4.6 MEDICAL AND HEALTH STATISTICS

The preceding sections of this chapter have already demonstrated some of the potentialities of applying statistical ideas and methods to medical problems: the handling of natural variation, the reduction of data, the testing of hypotheses, the use of efficient experimental designs, the investigation of dynamic processes, etc. These developments represent an enormous increase in the scope and power of scientific approaches to the management and interpretation of health data. They contrast strongly, however, with the image presented by the more traditional forms of vital statistics, mortality and morbidity records, and general public health data, which are often largely administrative in content.

These latter activities are easily conjured up by the terms 'medical statistics' and 'health statistics', and sometimes tend to project a somewhat boring bureaucratic image. In fact, the patient and accurate collection of such material on the national or international level can have important uses in drawing attention to significant trends in disease-rates, or in assessing the requirements for certain kinds of medical care. However, serious criticisms can be raised as to whether the cost and magnitude of the efforts involved justify, from a cost–benefit point of view, the results obtained.

Routinely collected data are apt to deteriorate in quality as time passes unless they are seen to be useful not only in principle but in actual practice. Motivation is a major requirement if data are to be amassed efficiently, accurately and without undue bias. Life insurance companies manage to work with vital and morbidity statistics in a fairly effective way because they would risk going out of business if they were careless with the data. But what is the motivation for producing good public health statistics, especially as the time for taking any effective action based on such material may be long past by the time publication is finally achieved? Moreover, the details of the information presented are usually too slight to enable any real epidemiological conclusions to be drawn as to possible causation. For these purposes specially designed *ad hoc* investigations are required. The latter can be carried out in depth over shorter periods of times, have a far higher scientific content, be specifically geared to answering explicit theoretical and practical questions, and are liable to be intrinsically more reliable because of the psychological interest and enthusiasm generated.

These considerations have become intensified in recent years by the increasing availability of electronic computing facilities. Work which was previously extremely laborious can now be undertaken relatively easily. Many people, especially senior officials and those on committees, are believed by some to be impressed by the large numbers of statistical tables which the machines can generate. As a consequence, more and more effort is devoted, at least in certain quarters, to producing vast quantities of computer print-outs that nobody has time to read or really wants anyway. The point has even been reached when *ad hoc* surveys are planned on a grand scale because computerization will easily produce a compelling stack of tables!

These matters will be pursued in more detail in Chapter 6, which deals with the use and abuse of computers. For the moment we merely note that thoughtless use of mechanization has forced the collection of much routine statistical data into a *reductio ad absurdum*. If medical and health statistics are to play a significant part in the current drive to improve health they must be much more closely connected with some kind of practical applications, whether these entail a search for fundamental medical knowledge or an attempt to increase the effectiveness of public health decision-making.

In more enlightened establishments the systematic collection, storage, and use of data are under constant scrutiny. Highly structured 'data banks' are built up in the expectation that they will be useful in testing new ideas as research progresses. However, we must always remember that since ideas and hypotheses are by definition 'new', it is in general unlikely that they will be adequately catered for by past methods of data collection. Again, there are modern methods of increasing the information available to epidemiologists by relating diverse medical records that refer to a single individual: the technique of 'record linkage'. By properly combining material drawn from various sources, such as hospital records, general practitioner records, national health insurance records, etc., using computerized methods for the most part, one can set up a master file containing a wide range of interrelated information.

Because general health statistics are liable to be insufficiently detailed to make a real impact on either research or administration, it is common practice nowadays to establish specialized registries in key areas such as childrens' immunization, genetic counselling, blind persons, cancer, psychiatric cases, ischaemic heart disease, radiotherapy, occupational hazards, medical hazards, etc. These cover preventive medicine, specific diseases and disabilities, treatment, aftercare, special risks, and so on. If the registries are actively used for well-defined purposes they tend to maintain interest, can be moulded to changing circumstances, and can be abandoned if no longer needed without doing violence to some larger monolithic system.

One useful aspect of registries is that they can be seen as providing opportunities for monitoring and surveillance in special areas. This allows investigation of the changing patterns of disease and medical care and assists the evaluation and planning of health services. There are many statistical methods, developed originally in the more precise area of industrial quality control, that can be adapted to the problems of monitoring. Health data are inevitably somewhat vaguer, and are liable to have strong anecdotal components when we are, for example, trying to associate apparent increases in certain disease-rates with specific environmental, behavioural, or nutritional factors. Nevertheless, the important thing is to be able to generate immediate alarm signals whenever there is a reasonable likelihood of a real effect existing, so that we can investigate the situation more fully. The availability of reliable registries increases the chances of linking cause and effect at an early date so that appropriate public health action can be undertaken.

There has been a concerted effort in recent years to see many, if not all, of the foregoing activities as part of an all-encompassing 'information system'. Whether there should be, or indeed can be, a single information system within any large undertaking or organization seems to be a debatable issue. The concept of a number of appropriately designed information systems, all suitably linked, appears to be more viable. We are, of course, using the term 'information system' in the present context to mean, in a rather general way, information that is not merely raw data but which has been organized in some deliberate manner to be of direct assistance to managers and decision makers.

It is immediately obvious that the types of information required by, say, a minister of health, a regional health administrator, a hospital director, a consultant in charge of a clinic, a general practitioner, etc., are all different, though strongly related from a health point of view. But many of the activities are partially supported by purely administrative processes, involving personnel, pay-roll, supplies, office maintenance, etc. These also depend on efficient information services, but can and should be handled quite independently.

Medical and health statistics can thus be taken to cover a large spectrum. In practice, most people would not include administrative and 'housekeeping' functions, but would be very likely to incorporate anything with an appreciable medical or public health content. In order to avoid any risk of health statistics losing the vigour necessary to solve modern problems it is essential that it should make full use of contemporary statistical techniques and maintain close contacts with several cognate disciplines to be discussed later, such as operational research, systems analysis, modelling, and computer science.

4.7 MEDICAL DIAGNOSIS

Let us now consider the more detailed application of probability and statistics to one special area in medicine by way of illustrating the power of this approach.

Few would deny that the whole process of diagnosis is of central importance in medicine, and that the process demands, moreover, a high degree of skill, knowledge and imagination in its practitioners. It is also universally recognized that the accuracy of a diagnostic decision and the speed with which it can be reached vary enormously according to the patient's condition; the data available on his individual symptoms, signs, and laboratory tests; the general body of medical information about the occurrence of such observable material over a wide range of alternative diseases; and the ability of the clinician himself. Moreover, the medical treatment adopted and the patient's expectation of recovery frequently depend on early and accurate diagnosis. With such tremendous variations in the effective efficiency of diagnostic procedures it is quite natural to try to determine the circumstances under which optimal results can be achieved.

Physicians have, of course, been making this attempt, with different degrees of success, for centuries. But with modern methods of diagnosis and treatment, using all the special skills of science and technology, the potentialities for success have been greatly enhanced in recent years. It is therefore important to have precise methods for discussing, investigating, evaluating, and controlling the process of diagnosis. As we have repeatedly claimed, the best way of achieving precise logical thought is by a mathematical approach. This approach can, in principle, be adopted no matter how difficult and complicated the subject. Indeed, if there are a large number of interdependent factors all showing appreciable natural variation, it is *only* by using an appropriate statistical method that the complex pattern of correlated effects can be handled with a reasonable degree of efficiency. If the numbers of factors or items of data are very large it may be desirable, or even necessary, to use a computer in order that the required results can be obtained in a sufficiently short period of time. This attitude by no means depreciates the value of insight and imagination. On the contrary, it allows these faculties greater scope by arranging for all tasks that *can* be couched in numerical and logical terms to be handled by mathematical and computer techniques.

What, therefore, can be done to apply these notions to the realm of medical diagnosis? There are already, as is well known, several enthusiasts for the use of mathematics and computers in this area amongst the ranks of mathematicians, statisticians, computer experts, and clinicians. The sympathies of the writer are naturally with the enthusiasts. At the same time, it is essential to have a clear idea of what theoretically could be achieved, what in the light of practical circumstances should be attempted, and what the implications are for the whole patient–doctor relationship.

Even if we never wanted to use computers for diagnosis in practice, there would still be a *prima facie* case for undertaking a mathematical analysis of the processes involved because of the likelihood of obtaining fresh understanding and insight. In fact, all the evidence to date suggests that there is a very real role for the computer in medical diagnosis. The primary task is to determine exactly what the boundaries of that role are, and we shall discuss various details in the sequel. For the moment we merely observe that some of the relevant aspects are the actual calculation of a differential diagnosis in appropriate circumstances, the analysis of diagnostic skills and performance with a view to the general improvement of standards, the construction of teaching aids for students, and the whole business of collecting, summarizing, and processing clinical data for expert consideration by human diagnosticians.

We shall be mainly concerned in the present chapter with an introductory account of how mathematical models of certain aspects of the diagnostic process can be constructed, and how these can be used to aid clinicians in evaluating at least some kinds of complex and highly variable medical data. As yet the whole subject of computer-assisted diagnosis is very much in its infancy, but sufficiently encouraging results have already been obtained by research workers in several different countries to make further investigation and development eminently worthwhile.

Of course, the concentration on the choice of a differential diagnosis is in many ways an oversimplification of the total problem, or at any rate a restricted view. We shall assume that the alternative diagnoses, between which a choice is to be made, are all clearly and unambiguously defined. In practice this is often not the case. Expert opinion is frequently divided about the best ways of classifying diseases (in contradistinction to allotting patients to predetermined classifications), and new knowledge may easily demand a revision of existing schemes. This naturally involves questions of medical taxonomy, and we might, on a wide view, want to consider the implications of a numerical taxonomic approach.

Again, treatment of an individual patient depends in large measure on the preliminary diagnosis. This diagnosis may be revised if the patient does not respond to the treatment that was thought to be best, or reacts in an unexpected way. Response to treatment may in fact be regarded as a late-stage diagnostic test, and as such is a fresh source of information. This is, of course, a clinical commonplace. But the point being made here is that we may well require a mathematical description of the whole process of disease classification, differential diagnosis, and response to treatment before any substantial advances can be achieved with this kind of approach.

The literature contains a fairly large number of scattered papers on the subject under discussion, but a really authoritative work of reference has yet to be written. For a general coverage of a wide range of topics connected with medical diagnosis the very interesting University of Michigan conference report *The Diagnostic Process* (Ed. J. A. Jacquez, 1964) is well worth reading. Several relevant contributions have also appeared in the series of Rochester Conferences edited by K. Enslein (1962, 1964a, b, 1966, 1968). We shall not attempt to provide a comprehensive list of references, but shall, in addition to the above books, quote a small number of important or convenient papers to which further reference can usefully be made.

Elementary discrimination between two diseases

Let us consider, to begin with, the simplest problem of discriminating between two diseases, or between two substantially different clinical conditions. We shall suppose that there is a standard check-list of symptoms, signs, laboratory tests, items of medical history, etc., which is routinely applied to the patients in question, all of whom are assumed to have one or other of the two diseases. Sooner or later a firm diagnosis is made. Therefore, so far as existing records are concerned, we have bodies of observational data relating to each of the two diseases. When a new patient appears for diagnosis the physician in charge of the case reviews the available records in some appropriate manner and allots the patient to the disease that seems most probable. This reviewing of the records is likely in general to be based largely on memory, though numerical summaries of the data might be referred to for confirmation of first impressions depending on intuition and experience. The question is, what can or should be done if a precise mathematical evaluation of the situation were to be attempted, extracting and using the maximum amount of statistical information contained in the records?

We shall examine this question in more detail in the next section, considering in fact the wider problem of discriminating between several diseases. In the present section we shall look at a simple, though highly successful, numerical procedure for two diseases that incorporates some of the ideas of statistical discriminant analysis. The latter technique classifies objects into one of two groups on the basis of a *score*, which is calculated as the sum of a number of suitably weighted components, each of which measures in some way a particular characteristic. If the score for a given object is above a certain threshold value the object is assigned to the first group, say, and if the score is below that value the object is put in the second group. The process is not, of course, infallible and mistakes may occur. But it is possible to determine optimal values for the weighting factors and threshold such that misclassification is reduced to a minimum.

Crooks, Murray, and Wayne (1959) have used such an approach for the diagnosis of thyrotoxicosis. While, in general, this is not clinically difficult, there are often cases in which the correct decision is in doubt, different observers having different opinions or the same observer changing his mind on different days. It was therefore decided to develop a simple numerical evaluation of each new case to see if any improvement in diagnosis could be obtained, especially in the doubtful cases. Table 4.1 shows the list of signs and symptoms to which special attention was paid, the rule for inclusion being that there had been recent onset or increased severity. The table gives a numerical score for each sign or sympton, according to whether it was present or absent on the basis of the criterion mentioned.

The scores allotted to the individual signs and symptoms were originally developed not from a sophisticated statistical investigation but from a more or less intuitive and empirical appraisal of existing records. Characteristics thought to be of greater diagnostic importance were given larger numerical values, the classifications being worded so that 'presence' was always indicated by some positive number and 'absence' by some negative number (zeros were used to indicate no significance either way). Certain modifications were then introduced, such as reducing very high values to diminish the consequences of variations between different observers.

The scoring procedure was then applied to two standard groups of individuals, one containing 83 definitely thyrotoxic cases and one containing 99 nontoxic persons, some of whom were quite normal and healthy and some of whom were patients with various conditions including simple goitre. A total score was calculated for every individual according to the agreed rules. There was naturally considerable variation in total scores, but the important point was that the toxic group produced a set of numbers that were all higher than the corresponding set for the nontoxic group. New modifications in

Table 4.1 Weighting factors for signs and symptoms in suspected thyrotoxicosis

	Present	Absent
Dispnoea on effort	+ 1	0
Palpitations	+ 2	0
Tiredness	+ 2	0
Preference for heat	0	− 5
Preference for cold	+ 5	0
Indifference to temperature	0	0
Excessive sweating	+ 3	0
Nervousness	+ 2	0
Appetite increased	+ 3	0
Appetite decreased	0	− 3
Weight increased	0	− 3
Weight decreased	+ 3	0
Palpable thyroid	+ 3	− 3
Bruit over thyroid	+ 2	− 2
Exophthalmos	+ 2	0
Lid retraction	+ 2	0
Lid lag	+ 1	0
Hyperkinetic movements	+ 4	− 2
Five-finger tremor	+ 1	0
Hands		
Hot	+ 2	− 2
Moist	+ 1	− 1
Casual pulse rate		
Auricular fibrillation	+ 4	0
Regular rate		
Over 90	0	− 3
80 to 90	0	0
Under 80	+ 3	0

the weighting factors were now introduced to produce the widest possible separation between the two groups without unduly emphasizing the previously mentioned observer effects. The final weighting factors for the individual signs and symptoms are those actually shown in Table 4.1.

Using these modified weighting factors, the total scores for the definitely thyrotoxic cases varied between + 21 and + 42, while the corresponding range for the nontoxic group was − 16 to + 10. All persons with scores of + 21 or more were thus thyrotoxic, and all those with scores of + 10 or less were not. A very good separation of the two groups is therefore obtained on the basis of the total scores. Although there were no individuals here who did not fall quite clearly into one group or the other, we should have expected some ambiguity if the sizes of the samples had been much larger. However, the acid test for an empirical study of this kind is its performance in relation to new data. It is always possible to invent predictive rules of apparently phenomenal accuracy by using sufficient ingenuity in manipulating a given body of data. The proposed criteria were accordingly validated by application to an entirely new group of individuals.

It would be interesting to use these scoring methods on new individuals, some of whom were thyrotoxic and some of whom were not, in order to see whether the rules specially devised for the original groups would still hold good. For a careful mathematical statistical study one would certainly want to do this. But Crooks, Murray, and Wayne chose a rather more heroic test. They selected 118 cases, all of whom had already presented diagnostic difficulties to hospital physicians. That is, they did not take cases in which clinical judgement had found little trouble in reaching satisfactory conclusions, but concentrated on doubtful cases, whose final diagnosis was determined only after prolonged observation and response to treatment. There were 51 cases eventually diagnosed as toxic. Of these 88 per cent had scores of $+20$ or more, while the remaining 12 per cent fell into the equivocal range between $+11$ and $+19$. Amongst the 67 cases finally found to be nontoxic, there were 88 per cent with scores of $+10$ or less and about 10·5 per cent in the equivocal range $+11$ to $+19$, with a single case (1·5 per cent) erroneously allocated to the toxic range. Similar diagnostic accuracies, averaging about 85 per cent, were also obtained in evaluating the status of 171 patients at four other centres.

It is remarkable that such a simple mathematical method was found capable of correct discrimination between toxic and nontoxic patients in precisely those instances where the clinical assessment had run into difficulties. There is obviously little point in routinely carrying out a computational technique in circumstances where human judgement is rapid, accurate, and readily available. But if the technique can be used to assist the formation of reliable conclusions in situations where expert opinion is uncertain, then it evidently has an appreciable practical significance.

The foregoing numerical method could be made somewhat sharper by undertaking a full statistical analysis using a discriminant function technique. We should expect rather similar conclusions, but with improved accuracies. As the discriminant function approach is difficult to generalize satisfactorily we shall not discuss the mathematical details here, but rather allow this subsection to demonstrate how much can be done without recourse to highbrow procedures.

The general case of several diseases

In general, it is frequently necessary to discriminate between several alternative possible diseases. It would probably be extremely troublesome to obtain a satisfactory empirical scoring system along the lines just described. Moreover, the absence of a well-defined underlying mathematical model would lead to many difficulties in the interpretation and evaluation of results, and it would be hard to develop new ideas. We could, of course, try using the mathematical discriminant function approach referred to above, though this again raises several problems. In the first place, the classical treatment for discriminating between two groups assumes that the covariance matrices of the symptoms measured are the same in each group, and that the distributions are multivariate normal. The former assumption is not really essential but leads to a fairly simple mathematical analysis. A more general treatment with unequal covariance matrices would be quite feasible, though almost certainly requiring a good deal of computing. Again, the assumption of multivariate normality is unlikely to be generally acceptable. Even more serious is the difficulty of extending the discriminant approach to several groups (unless these can be graded in some unique linear order, and the assumptions of multivariate normality and equal covariance matrices hold). Use could be made of Mahalanobis' generalized distance, but there would still be many complications in the application of the theory and the interpretation of results.

There are actually quite a number of statistical approaches that can be tried, and Masuyama (1963) lists a total of eleven methods, though only three or four of these are substantially different. However, one or two particular procedures seem to the writer to have considerable advantages. One involves developing an analysis in terms of the probabilities of occurrence of individual symptoms for particular diseases, and leads to the calculation of the probabilities that a given patient has any specific disease in the group to be discriminated. (See, for example, the work of Ledley and Lusted, 1961.) The method, in fact, involves a Bayesian approach (see Section 4.3). Since the basic ideas are couched in terms of simple probabilities, the underlying mathematical model is easy to describe, and the meaning of practical results is readily understood. Let us therefore examine this approach in more detail.

Suppose that we are concerned with a limited group of d alternative diseases indicated by $D_1, D_2, \ldots, D_d$, and that any given patient is to be diagnosed as having just one of these diseases (i.e. the complication of multiple diagnosis is excluded from the present simplified discussion). We also assume that there is a specific list of s signs, symptoms or laboratory tests, on which information is normally elicited, indicated by $S_1, S_2, \ldots, S_s$. It is often convenient to refer to a whole set of signs, symptoms, etc., for which purpose we shall use the symbol S without a suffix. For the time being we shall assume that each symptom is discrete, i.e. classifiable into two or more specific classes (like 'present' and 'absent' for simple dichotomies).

Now, suppose that there are extensive past records on patients in all the disease groups so that, for example, *given* any disease D_j, we can specify the probability $P(S|D_j)$ of a particular symptom complex S being observed. Further, if the records are a representative collection we shall also be able to determine the probability $P(D_j)$ that a patient selected at random has disease D_j, that is before any clinical information on that particular patient is taken into account. The quantity $P(D_j)$ is, of course, the *prior probability* of disease D_j. Although an examining physician may well not think explicitly in terms of such a probability, he is quite likely to give more attention to the possibility of the patient's having a disease known to be common rather than one known to be rare.

The next stage in the process is for the physician to collect information about the individual patient's symptoms. Quite probably a tentative diagnosis will be made very early on, often with certain alternatives in mind. There comes a point, however, when at least for the time being as much relevant data as possible have been collected. These data define a particular symptom complex S. For the purposes of immediate treatment and perhaps a request for further special tests, a tentative diagnostic decision must be made. In probability terms the physician wants to know the probability attaching to each disease, *given* the symptoms, i.e. $P(D_j|S)$. This is really only a numerical way of describing what happens in clinical practice. If a very firm diagnosis is made in favour of disease D_1, the probability $P(D_1|S)$ will be relatively high, all the other probabilities being very much smaller (perhaps by a factor of 10 or 100). If, on the other hand, two diseases, D_1 and D_2, seem about equally likely and all others are excluded, we should have $P(D_1|S)$ and $P(D_2|S)$ each about 0·5 and all the rest near zero. And so on for other situations that might arise. The physician, of course, normally bases the degree of belief he has in any diagnosis on his general experience and judgement. But ultimately these must always be traceable to his knowledge of previously collected or observed clinical data (either his own or that of others presented in medical journals and textbooks) and what he can deduce from the patient under examination.

It is well known that the *posterior probability* $P(D_j|S)$ is proportional to the product of the prior probability $P(D_j)$ and the *likelihood* $P(S|D_j)$, i.e.

$$P(D_j|S) \propto P(D_j)P(S|D_j). \tag{4.11}$$

Since the sum of all the posterior probabilities for the different diseases must be unity, we can write (4.11) more exactly as

$$P(D_j|S) = \frac{P(D_j)P(S|D_j)}{\sum_{j=1}^{d} P(D_j)P(S|D_j)} \quad (j = 1, 2, \ldots, d). \tag{4.12}$$

This is essentially Bayes' rule for calculating the probability of an hypothesis, given its prior probability and some observed data. The method is not strictly valid unless the alternative hypotheses (in this case diseases) have a prior probability distribution. It is, however, always open to us to adopt an inferential model (see Section 4.3) in which the prior probabilities do exist, and are specified by appropriate rules even in the case of complete ignorance.

The computation of the posterior probability of any disease is in principle quite simple, since both $P(D_j)$ and $P(S|D_j)$ should be available from the existing records. But care is needed with the likelihood $P(S|D_j)$, since even with very extensive records there may be very little information about the frequency of occurrence of a specific symptom complex S. For if there are s different symptoms and they all are dichotomies, there will be 2^s separate symptom complexes. If $s = 20$ then $2^s > 10^6$, an unmanageably large number. Some simplification is ordinarily required, and one frequently made is that the individual symptoms are independently distributed. This supposition is possibly not as drastic as might be supposed. Thus van Woerkom and Brodman (1961) concluded that appreciable correlations between different symptoms were not present in their material to any great extent. However, such assumptions are certainly not always justified. As shown by Norusis and Jacquez (1975), symptom nonindependence can be appreciable, and requires special methods to take it into account.

If the s symptoms are taken to be statistically independent of one another we can write

$$P(S|D_j) = \prod_{i=1}^{s} P(S_i|D_j) \quad (j = 1, 2, \ldots, d), \tag{4.13}$$

and the marginal probabilities $P(S_i|D_j)$ may well be calculable from the records to a good degree of accuracy.

Although for ease of exposition we have assumed that all the symptoms are discrete in form, this restriction can easily be relaxed to include continuous measurements. Of course, one way to deal with continuous variables is simply to define some range of variation as 'normal', and everything outside this range as 'abnormal'. We can then work with a simple dichotomy of 'present' and 'absent'. But this is to throw away a good deal of statistical information, and it is better to use the continuous distribution as such if at all possible. Suppose that the kth symptom is continuous with frequency function $f_{kj}(x)$ for the jth disease. This means that the probability of an observed measurement lying in the range x to $x + dx$ is $f_{kj}(x)dx$. Substitution of this probability in (4.13) leads to an element dx in $P(S|D_j)$. This element will also appear in the numerator and in every term of the denominator on the right-hand side of (4.12), and will therefore cancel out and cause no trouble. The extension to continuous variables thus involves no more than replacing the previously discrete probability by the corresponding continuous frequency function, calculated for the measurement in question.

The above probability description of the diagnostic process is obviously a rather simple one, and perhaps a little naïve in certain respects. It has nevertheless been surprisingly successful in a number of special applications, as we shall see shortly. For the moment let us consider one outstanding practical major criticism. This is that a physician is confronted with a real live patient, and may frequently, if not usually, draw all kinds of useful and relevant conclusions from small clues noticeable only to a trained observer, without his necessarily being explicitly and consciously aware of what is happening. The probability calculations, on the other hand, can deal only with carefully recorded data and may miss a great deal that is available to direct observation. Two comments are worth making on this objection. First, in certain circumstances the mathematical assessment may do as well, if not better, than a human diagnostician, in which case it is missing little of diagnostic value. Secondly, if computed diagnoses turn out to be markedly inferior then there is either some inadequacy in the mathematical method as such, or the information available for processing by that method is appreciably less than that usable by a human observer. In either case, identifying and rectifying the causes of the inefficiency will produce new knowledge about the diagnostic process and, quite possibly, a greatly improved computational technique.

One of the earliest and most interesting applications of the foregoing method was the work of Warner, Toronto, Veasey, and Stephenson (1961) on congenital heart disease. They had a list of 33 separate disease conditions and 50 distinct discrete symptoms. Thus the basic sympton–disease matrix summarizing the existing information on the probability of occurrence of each symptom for any disease contained 1650 items for a start. It is clear that the use of equations (4.12) and (4.13) above to evaluate the data from a new patient involves a considerable quantity of elementary, though tedious and time-consuming, arithmetic. The use of an electronic computer was therefore highly desirable if not absolutely essential. This raises an important point about the future use of computers in this context. As the number of observable symptoms increases, and especially as the number of laboratory tests expands, we may soon come to the point where there is more information than even an experienced diagnostician can conveniently handle. We are then faced with the paradox of being potentially able to achieve far more accurate results, but of being unable to do so in practice and in fact doing worse because of the embarrassingly large quantities of data (compare the wartime high-frequency direction-finding situation discussed in Section 7.2). The use of a computer to store and handle the basic data as well as carry out all the probability calculations would then become mandatory.

Warner *et al.* found that the use of the probability method was highly successful. Firm diagnoses were eventually made in practice by cardiac catheterization or actual surgery and could therefore be used to check the probable clinical diagnoses proposed in advance by cardiologists and by the computer. In a series of 36 cases it appeared that the most probable computed diagnosis agreed with the true diagnosis at least as often as the conclusions drawn from the same clinical data by three experienced cardiologists. These striking results were strong confirmation of what could be achieved by mathematics and computers in an appropriately chosen field of application.

Unfortunately, Warner *et al.* do not give a detailed statistical analysis of the comparisons between clinical and computed diagnoses. Subsequently, Gustafson *et al.* (in Enslein, 1964b) carried out some rather similar work on

Table 4.2 Comparison of diagnoses by computer and clinicians

		First choice correct	Second choice correct	Correct diagnosis otherwise mentioned	Correct diagnosis not mentioned	Score
Computer		63	16	18	3	2.39
Pediatric cardiologists	1	56	20	8	16	2.16
	2	74	5	8	13	2.39
Other clinicians	3	36	32	4	28	1.75
	4	52	4	18	26	1.96
	5	36	12	28	24	1.57
	6	48	0	4	48	1.48

congenital heart disease using a considerably modified version of the original computer program. These authors do give comparative figures for different clinicians. A series of 38 patients were examined and all symptoms made up into check-list form. Each check-list was then used to supply the computer with basic data for calculation. It was also presented to six clinicians, comprising two pediatric cardiologists, two internists with special interest in cardiology, a pediatrician, and a cardiac surgeon. The clinicians were asked to list all likely diagnoses for each patient in order of probability, while the computer program was arranged in such a way as to print out the four most probable diagnoses in numerical order. Results are summarized in Table 4.2, and are more or less self-explanatory. Thus the computer's first choice was correct 63 per cent of the time, its second choice was right 16 per cent of the time, and in 18 per cent of cases the proper diagnosis was at least mentioned in the alternatives provided. Only 3 per cent of correct diagnoses failed to be mentioned at all. A score was attached to each source of diagnosis by counting three points for a correct first choice, two for a second, and one for a mention.

Although it may be considered that the computer was fairly successful here, the authors emphasize that an experienced pediatric cardiologist working with his own material would make a correct diagnosis about 90 per cent of the time. His data would then include ECGs and X-ray material. These were not available in check-list form, so that neither the computer nor the six clinicians in the comparisons of Table 4.2 could make use of them. However, it was pointed out that although experts can handle this type of information, less well-trained clinicians cannot. One concludes from these results that even in the present state of the art the mathematical method will extract as much, if not more, information from a suitable set of quantitative records as most human experts. At the same time some of the latter have the advantage that they can utilize material that is so far difficult to feed into the computer. Obviously, further efforts are desirable to find ways of supplying the computer with the additional data.

Substantially similar results to the above have been obtained by several other workers in different parts of the world, mostly in the field of cardiology (see, for instance, Takahashi *et al.*, 1963; and Vishnevsky *et al.*, 1964) using either the foregoing Bayesian method or some other related statistical technique.

Useful applications have also been made to the identification of bacteriological specimens, where the basic principles are essentially the same as those we have been discussing (see Payne, 1963). Thus existing records show the frequencies with which standard tests yield specific results for a wide range of known bacteria. Given a new unidentified specimen, a selection of tests are performed and a tentative identification made. If this is not sufficiently firm more tests are undertaken until an acceptable diagnosis is reached. In current bacteriological practice this process is in general carried out on the basis of judgement and experience, and a good bacteriologist will achieve correct results quickly and efficiently. There are, however, more difficult situations in which the results of tests are rather variable. In such cases it may be helpful to compute a probability diagnosis on the basis of the preferred initial batch of tests. If sufficient discrimination is not obtained, it is possible to compute the most efficient batch of tests to be carried out next. Such ideas are equally important in the realm of diagnosing disease, when it will usually be desirable to reach a decision on a miniumum amount of investigation (especially if some tests are unpleasant or even dangerous).

We have really only scratched the surface of the potentialities of mathematical and computing methods in medical diagnosis. It is evident from the specialized application to the problem of thyrotoxicosis that very simple mathematical procedures can be of undisputed help in appropriate circumstances, and the more advanced techniques have already shown considerable promise at a fairly early stage of research and development. We may reasonably expect that further work will greatly increase the possibility of the rapid and accurate evaluation of diagnostic material.

If such methods are to be used routinely in suitably chosen areas, satisfactory solutions must be found to the problems, already touched upon, arising from the relative paucity and incompleteness of many available records, the possible correlation between different symptoms, the occurrence of multiple diagnoses, and the frequent desirability of starting some treatments before a firm diagnosis has been achieved. Several advances have been made in recent years, especially in handling the problem of correlated symptoms (see, for example, Anderson and Boyle, 1968; Scheinok, 1971; Davies, 1972; Norusis and Jacquez, 1975).

It is clearly important that real problems are investigated, and not purely mathematical abstractions. This raises the whole question of what mathematical models of the diagnostic process are really acceptable. The foregoing sections have entailed certain definite, though oversimplified, mathematical versions of what diagnosticians do or attempt to do in real life. Far more study is required of what actually happens in clinical practice, so that more realistic models can be constructed. As we have seen, mathematical computation based on simple ideas can often do as well if not better than a good diagnostician when both are working from the same check-list. But the diagnostician may well be able to do better when dealing with a live patient, because additional information is then presented to him that is not available to the computer. It may, however, be possible using modern methods of automatically recording and analysing clinical data (such as physiological records) or using the developing techniques of automatic pattern recognition (as in the counting and classifying of chromosomes) rapidly to tap vast sources of information that are otherwise accessible to the clinician only slowly and incompletely. If this comes about, the computer evaluation of diagnostic material may be able to achieve a speed and efficiency quite beyond anything realized so far.

What we really need is a concerted attack on all major aspects of the whole diagnostic process, with theoretical mathematical studies and computer programming methods closely integrated with actual clinical applications. Only if these researches can survive the acid test of continued operational validation is the initial promise of the work described in this section likely to achieve fulfilment.

CHAPTER 5

The Art of Modelling

5.1 WHAT IS A MODEL?

In a way this chapter should be unnecessary. We have already used the word 'model' many times in previous chapters in connexion with a scientific approach and the application of mathematical methods. The sense of the word has usually been some kind of picture or description of reality that can be used for further thought and discussion. However, this is not primarily a scientific or logical concept: it is a fact of everyday, commonsense usage. Why, then, is it necessary to discuss models and modelling as though they were some independent and specialized activity, understood only by the initiated?

A full answer to this question may not be possible at the present time, and might in any case take us too far afield. But it is clear that there are major historical and psychological reasons for singling out the entity 'model' for special consideration. The net result has been that the word is now surrounded by a jargonistic aura that makes people imagine almost insuperable difficulties where none exist. Some discussion is therefore worthwhile to restore the word to its proper semantic status, and to indicate the implications for a variety of quantitative approaches to health problems.

Referring back to the development of ideas about the motion of the sun, earth, moon, and planets in Section 2.1, we can say that all the different cosmological descriptions due to Aristarchus, Aristotle, Ptolemy, Copernicus, Kepler, and Newton constituted a series of models. The later models were sufficiently detailed to be considered as serious scientific hypotheses, and the increasing power to explain observed data represented a clear advance of quantitative knowledge. In this respect the construction of a model is an integral part of the scientific approach. The idea of a model probably raises few difficulties in anyone's mind when considering the dynamics of physical objects, whether in astronomy or in ballistics. The possibility of constructing a small-scale physical model, by way of illustration, somehow prevents the notion from seeming too remote. But when the phenomena of electromagnetism or quantum theory and relativity are being investigated, primitive physically constructible models quickly give way to systems of mathematical equations. The implicit modelling is then considerably more rarefied.

Similar difficulties arise in biology. In Section 3.4 we saw how various geometrical descriptions could be employed to account for the broad appearances and growth characteristics of shells. Although the underlying mathematics may be difficult, there is little difficulty in visualizing and actually drawing the various geometrical patterns that can be discerned. On the other hand, the stochastic processes of Section 4.5, used to describe in summary fashion some very complicated dynamic activities involving major statistical aspects, represent a considerable increase in the complexity of the modelling concept. Further examples of a sophisticated type of probability modelling occur in Section 4.7 on medical diagnosis.

Thus, one soon arrives at a point where the hypotheses, and the concomitant modelling, achieve a considerable degree of abstractness and complexity. Difficulty is then experienced in understanding the ideas proposed, let alone knowing whether they are reasonable and likely to promote new advances in understanding. Moreover, the handling of these approaches often seems to necessitate the application of advanced mathematics and computerized techniques. If one is not an expert in these matters, how can one judge their utility? When this situation is reached, which happens rather easily in medicine and public health, perhaps there is something to be said for going back to first principles, and asking, 'What exactly is a model?', 'What do they mean?', 'How are they to be tested and used?' and so on.

We can therefore conveniently see a model as a well-defined, well-organized picture or description of some aspect of the real world that we are specially interested in. A model can be mainly qualitative in character, though the scientific approach we are advocating is most effective when it has a large quantitative component. In this case the associated model will be also quantitative, and indeed mathematical, in form. Hence the frequent use of the term *mathematical model*.

It is important, however, to realize at this stage that, although a mathematical model is essentially a mathematical formulation of the problem or hypothesis in question, the broad structure of the model can usually be couched in easily understood nontechnical language. Thus the Newtonian description of a planet moving round the sun involves the former moving in an elliptical orbit with the latter situated at one of the foci, the planet and the sun being subject to a gravitational force of mutual attraction which is inversely proportional to the square of the distance between them. Little scientific knowledge is required to interpret this statement in general terms, but a precise quantitative formulation requires a mathematical equation needing technical skill for its solution. At the same time, this equation is the most convenient way of describing the details of the planetary motion, all macrophenomena, including Kepler's laws, being easily deducible from it.

As an epidemiological example, we may consider the simplest model of an epidemic with infection and recovery followed by immunity pictured in terms of two boxes. The first box contains those who are susceptible; the second box contains those who are infectious and circulating freely in the population. The numbers of susceptibles and infectious persons are to begin with thought of as continuous. New infections take place with the continuous transfer of individuals from the first to the second box. The rate at which this happens depends on the product of the numbers of infectious and susceptible individuals. Infectious persons leave the second box in a continuous way when they present symptoms, are removed from circulation and isolated until recovery. In this case the rate of removal of infectious persons is proportional to the total number of infectious persons. Such a model, involving well-defined rates of transfer with no probability aspects, is called *deterministic* (see also Section 3.4). In the present case the behaviour of the model can be completely described in terms of two differential equations that are relatively easy to handle mathematically.

The assumption of continuously variable numbers of susceptible and infectious persons is a reasonable approximation when the numbers are large and probability aspects are correspondingly small. But, if numbers are small, we need a more realistic model taking probability factors into account. Such a model is called *probabilistic* or *stochastic* (see also Section 4.5). We may now assume that infections take place by random transfer of individuals from the first to the second box. The probability of transfer in any short interval of time can be taken to be proportional to the product of the numbers of infectious and susceptible individuals. Similarly, infectious persons leave the second box in a random way, the probability of a single removal in a short period of time being proportional to the total number of infectious persons.

Although the structure of this model is still quite simple, a precise mathematical formulation entails a whole set of simultaneous differential equations. An alternative mathematical formulation can be used to replace the set of equations by a single, but more complicated equation (in fact, a second-order partial differential equation with three independent variables). In either case considerable mathematical skill is required to deduce results, and over the years a large literature has arisen from this model alone. More realistic models are, inevitably, even more difficult to handle. Nevertheless, the equations in question do allow the behaviour of the epidemic to be investigated in great detail.

These examples provide a very good illustration of the way in which the broad description of the structure of a model, usually presented in the form of a simple diagram, acts as a link or *interface* between the complexities of the real world and our oversimplified, but quantitatively precise, representation of it. The structural diagram is often referred to as *the* model, and can usually be understood in a straightforward way both by the mathematical specialists who may be involved in manipulating the more precise mathematical version of the diagram and by the nonmathematicians who are faced with the real-life problem.

The importance of the latter aspect can hardly be over-emphasized. It means that we can draw on the whole vast field of mathematics to study and solve the equations that arise. Moreover, any mathematical conclusions that are reached, in the form of general theorems or more specific results, can always, via the model, be translated back into ordinary commonsense language.

Modelling can thus be seen as an integral part of the scientific method, in which mathematical analysis provides a more searching logical investigation of the implications of proposed hypotheses. This logical investigation, in its turn, leads one to deduce a greater wealth of practical consequences, permitting the hypotheses to be more stringently tested against observable facts. Well-tested hypotheses accumulate to establish well-tested theories. And well-tested theories lead to both greater understanding and improved control of processes in the real world.

5.2 TYPES OF MODEL

Having discussed at some length what in essence, a model really is, we now go on to consider the different types of model that exist, especially those of primary relevance to medicine and health. No attempt will be made to give a rigid and exhaustive classification; rather, a review will be presented of some of the major aspects of modelling which are likely to be useful in evaluating other people's modelling work or in developing one's own, perhaps in collaboration with more technically oriented experts who will deal with the mathematical and computing details.

Deductive and inductive models

To begin with, it is worth noticing that the scientific method as described in Chapter 2 involves distinct, though related, deductive and inductive phases. Putting it rather simply, deductive reasoning entails logical argument on the basis of assumptions (axioms, premises) that we are prepared to accept as true, while inductive reasoning is an attempt to argue from facts and data back to explanatory hypotheses. The associated modelling activities usually reflect these two aspects rather clearly, but the emphasis given to each depends on circumstances.

The relatively simple probabilistic model of an epidemic outlined in the previous section is mainly deductive in form. Simplifying assumptions are made that incorporate the major epidemiological features of many natural epidemic processes, but realistic refinements are excluded. If it is agreed that such a model is all right, so far as it goes in representing broad aspects, then we can deduce logically and mathematically what the implications are. In terms of hypothetical population size, initial number of infectious individuals, infection-rate and removal-rate, we can calculate the statistical distribution of the final total size of the ensuing outbreak. The ratio of the removal-rate to the infection-rate, often called the relative removal-rate, is a key index. Suppose, for example, that the initial number of susceptibles is n, the initial number of infectious individuals is a, and the relative removal-rate is ρ. Then it can be shown that the chance of a substantial epidemic building up is negligible if ρ is greater than of equal to n. However, if ρ is less than n, the chance of a minor outbreak is $(\rho/n)^a$ and of a major epidemic is $1-(\rho/n)^a$.

Results of this kind are of course rather theoretical, but they do provide some insight into the mechanisms at work. Moreover, the notion of a threshold level of susceptibles, below which a serious outbreak is unlikely, is a common idea in communicable disease epidemiology and has obvious implications for public health control. Of course, many other more sophisticated quantitative conclusions can also be deduced.

The kind of deductive discussion has now been developed to cover a large number of situations in communicable disease theory. The advantages are that we can elucidate the quantitative implications of purely verbal thinking and argument, unrealized contradictions can be made explicit, further hypotheses can be generated and tested, guidelines for possible control procedures can be suggested, and so on. These ideas are naturally of very general applicability.

Of course, everything ultimately depends on whether we can believe in the models. Their uses in clarifying thought and generating hypotheses are important, but not enough. The models need to be tested in the real world, and discarded or modified if they are inadequate. This is not always easy to do. Thus to describe in a statistically satisfactory way the pattern of occurrence of measles in small family outbreaks requires a comparatively complicated model. It is necessary to take into account a variable latent period, an extended period of infectiousness, and an infection-rate, requiring at least four parameters in all. Suffice it to say here that an appropriate model can be fitted to actual data, and estimates of the parameters obtained, together with the relevant standard errors. This has been done both for measles and infective hepatitis, for example.

As soon as we start fitting models to data, and changing or rejecting them if they are unsatisfactory, we have switched over to an inductive mode of operation, since it is then the data that are accepted and the hypotheses that we are trying to arrive at.

Deductive modelling could be used far more often in clarifying thought and working out the consequences of agreed assumptions. Thus, when a committee is trying to decide upon recommendations, there can be quite violent disagreements. If there is time for careful analysis it may appear that the contestants are starting from different assumptions, so that with impeccable logic they may well arrive at different conclusions. Sometimes conclusions are insensitive to changes in certain assumptions. This kind of knowledge can also be useful, as when one committee member claims that an argument or conclusion will be totally invalidated if an assumption is found later to hold only approximately, and not to be precisely true.

Again, a great deal of administration and planning uses deductive arguments. A decision maker cannot undertake extensive research to determine the best decision to be made now, in an hour's time, or even tomorrow. Given all the facts and information available now, along with the best explanations current, he wants to know what the implications are. He also wants to know, given all the assumptions, what would happen if he changed by administrative action some key factor.

While action usually cannot wait for research, it is important that an appropriate process of evaluation be undertaken. And this *will* entail the carrying out of scientific work designed to support the administrative actions in question. Incorrect or inefficient interventions can then be identified, admittedly after the event, but criticisms can be made in relation to a developing body of insight and knowledge. Future decisions can accordingly be based on an improved understanding.

Another related area relying heavily on deductive argument is the handling of large systems. We shall mention these again later in this section, and also more extensively in Chapter 8. For the moment we merely note that the system dynamics of Forrester (1961, 1969, 1971), and the derived world models promoted by the Club of Rome, are very much of this type. The deductive phase has been very intensively explored and a new awareness of world problems has emerged. Nevertheless, there has been a great deal of criticism, and it is clear that much more work needs to be done on the inductive side, testing the models and fitting them to data. In this way a body of knowledge will be built up involving assumptions that can be broadly accepted.

It could also be mentioned here that econometric models of socio-economic activities have been developing rapidly in recent years. On the whole this subject is outside our present coverage, but there is some overlap since health economics is an important discipline for public health. To begin with, econometric models were very much of the *a priori*, deductive kind, providing theoretical interest but no sure practical applications. Nowadays, though, many of them have arrived at the point where they fit observed economic data convincingly enough for them to be used reliably for at least short-term predictions, and thus be of considerable practical value of banks and other commercial enterprises. Again, we see the importance of a proper relationship between the deductive and inductive aspects.

Functional and structural models

Another way of looking at the position of modelling in medicine and public health is to relate it to the current debate in epistemology between 'positivists' and 'structuralists' (see, for example, Gardner, 1973). The positivist view stems from David Hume and the French Encyclopaedists, and maintains that experience is our only source of knowledge: empirical science is thus our only method for understanding the world. There is, of course, the well-known difficulty that the validity of the inductive reasoning thus involved cannot itself be logically proved nor derived from experience. Nevertheless, scientists since the time of Hume have generally adopted a positivistic point of view.

Positivism has clearly had an important influence on scientific thinking, especially in its insistence on observed facts rather than mere speculation. This was particularly helpful in the development of the human sciences, e.g. the foundation of psychology, ethnology, and linguistics in the nineteenth century. Later, however, the positivistic approach became constrictive since it tended to confine studies of human behaviour to mere taxonomic descriptions of observed data plus immediate inductive inferences. Real explanatory power was lacking.

On the whole, the physical sciences managed in practice to avoid this difficulty. For example, the positivist rejection of the atomic theory in the late nineteenth century, on the grounds that no one had 'seen' an atom, clearly had little influence on fundamental work in progress.

Structuralism, on the other hand, admits the possibility of innate knowledge, and seeks to explain the observed data of behaviour, or 'surface structures', by underlying relationships, or 'deep structures', that cannot be directly observed. A major difficulty with the structuralist approach in the human sciences is that, unlike physics and biology, the theories are hard to test by critical experiment or observation.

This is not the place to get into deep philosophical discussion. Nevertheless, it is interesting to note the differences of attitude between those workers who have adopted a positivistic or functional point of view, and those who have preferred an essentially structural attitude. Thus in psychology the pioneer work of Freud was clearly structuralist, providing a new approach that used underlying psychological mechanisms to explain observable surface behaviour. Verification of the basic assumptions has, however, been a constant source of difficulty.

In ethnology the work of Franz Boas was plainly functionalist, while that of Claude Lévi-Strauss is essentially structuralist. Linguistics provides examples of the positivistic approach in the school of Ferdinand de Saussure and Leonard Bloomfield, while Noam Chomsky's transformational generative grammer is a structural concept.

Examples could be multiplied indefinitely. The point is that, as frequently happens, there are two divergent, dialectically opposed trends, neither of which by itself is really adequate. Efforts to find an appropriate synthesis may lead to new developments. This situation is readily discernible in medical statistics where, as we have seen in Section 4.6, the traditional activities have been largely concerned with the surface observations of mortality and morbidity data. This always gives a good start to a developing science, but sooner or later will be found to have too little explanatory potential.

Thus the medical research workers, referred to in Sections 4.6 and 6.6, who rely on vast collections of sociomedical data hoping that computerized statistical analysis will automatically constitute worthwhile research, are heavily biased towards the purely functional approach. On the other hand, many theoretical model builders in epidemiology and econometrics go too far in the opposite direction, constructing explanatory mechanisms which are too complicated to be verified.

As usual, the scientific approach outlined in Section 2.1 is the best guide to resolving such dilemmas. Preliminary investigations of the natural history type are liable to entail extensive data collection followed by some systems of classification and analysis. To begin with, one summarizes various totals, rates, percentages, etc. One then looks for interesting variations and correlations between different factors or characteristics. With luck, some kind of quantitative picture begins to emerge. A great deal of noncommunicable disease research starts in this way. Of course, if there are large numbers of interrelated factors it may be very difficult to disentangle any manageable scheme of possible causation. Sometimes, as with smoking and lung cancer, one is fortunate in identifying a major factor, though minor factors may still have some importance. As is well known, correlation is not causation, though it may usefully suggest possible causal linkages.

The next step is to try to marshal the statistical data into some kind of helpful pattern. At this stage we are still probably dealing with surface phenomena, and are therefore very much in the functionalist mode of operation. If the statistical modelling, probably using some kind of multivariate analysis, is well chosen we shall not only see helpful ideas and hypotheses beginning to emerge, but may to some extent be able to validate the statistical assumptions by generating predictions that can be verified on fresh data. This kind of work can be found on an extensive scale in the epidemiological literature. Its explanatory power is liable to be low unless unifying mechanisms can be found which integrate a large number of otherwise isolated facts into a single model.

Consider, for example, the contrast afforded by communicable disease studies. Here, an underlying unifying process is frequently available in the biological mechanisms known to be at work. Even if this is partly qualitative it is usually not difficult to design a quantitative model that can be critically tested against observed data. Big advances have been made in recent years in the modelling of communicable diseases (see the review by Bailey, 1975). Although there is much of purely mathematical interest, a point has been reached for several diseases where useful results can be obtained that are of direct value to public health control. The population dynamic models for communicable disease spread are accordingly structural models and of great potential significance.

In a few cases, e.g. measles, infective hepatitis, and malaria, models have been developed and actually fitted in a proper statistical manner to specific data from given communities. But in the majority of cases, including tuberculosis, typhoid fever, cerebrospinal meningitis, cholera, tetanus, etc., only hypothetical situations have been investigated. That is, models acceptable to epidemiologists have been constructed, but, owing to lack of data, parameters have had to be estimated from heterogeneous sources, or even intelligently guessed. The models thus apply to typical situations, e.g. the spread of tetanus in a developing country, but are hard to apply to a specific community. For in the latter case a local administrator can always claim that conditions in his community differ in some essential respect from the assumptions incorporated in the model, such as that the local infection-rate is appreciably higher or the local immunity patterns are quite different.

The general models are nevertheless useful for investigating the possible consequences of certain forms of public health intervention, and can be helpful when dealing with problems of policy and choice of broad strategies. It is worth noting that, as soon as real-life problems have to be tackled, the whole modelling situation is thrown into much sharper focus. In the communicable disease field it is relatively easy to provide a critique that allows one to distinguish work that deals primarily with mathematical refinements from investigations that are progressing towards a practical instrument of control. Much of this is still in the future, but concerted efforts on parasitic diseases (see Chapter 1, for example) like malaria, schistosomiasis, hookworm disease, onchocerciasis, etc., could bring tremendous benefits.

To return for a moment to the noncommunicable diseases, one sees little prospect of the same kind of modelling. Of course, if we had better models of cancer growth, covering many different types of tumours and successfully validated by fitting to appropriate observational data, then there might be a considerable increase in our quantitative knowledge of the pathological processes involved. This in its turn would facilitate the evaluation of diagnosis, prognosis, and treatment. Again, it might be that improved understanding of an operational kind would be obtained if models could be constructed that permitted the prediction of health status in relation to factors under public health control. These possibilities will be discussed in greater detail in Chapter 8 when we come to look at the problems of large complex systems which need special kinds of approaches.

The message of this section is therefore that it is important to know what kind of modelling we are thinking of undertaking, or what is being proposed by other investigators. Confusion in these matters can all too easily lead to a concentration on inappropriate technical methods, with a resultant waste of time and money, not to mention getting a powerful method of investigation into disrepute. While the philosophical aspects discussed here are of major importance, no attempt has been made to achieve an exhaustive classification. There are may different ways of looking at models, depending partly on external circumstances and partly on the investigator's personal style. Some of the more important types of practical applications will be discussed in the following section.

5.3 LEVELS OF APPLICATION

We have already noted in Section 3.3 that quantitative and scientific approaches can be applied at many different levels of human activity. This is, moreover, equally true for the medical and health field (e.g. Sections 2.2 and 3.5), from individual small-scale technical activities at the bottom of the logical hierarchy to broad policy-making at the top. Mathematics, statistics, and computers may be useful at any level. And, as we have seen, all serious quantitative work is liable to involve some element of modelling. Indeed, modelling is implicit in most scientific work. In the present chapter we are, therefore, not so much concerned with the simpler applications already mentioned in Chapters 3 and 4, for example, but rather with a careful and deliberate attempt to make sense out of some more or less complex situation.

This is now coming very close to the subject matter covered by operational research and systems analysis, which are however specifically oriented towards operational and decision-making situations. Those will be discussed in more detail in Chapters 7 and 8.

In the present chapter we are more interested (as in the two previous sections) in what models are, what major logical distinctions have to be made between them, and what modelling activities involve when they are applied to very different kinds of problems at different levels. For the latter purpose it is convenient to identify certain broad classes of activity, such as the following:

(*a*) Technical models

These deal primarily with a scientific study of phenomena on processes in well-defined technical areas. At the level of the individual there is already an enormous literature covering a very wide range of quantitative modelling. For example, the double-helix structure of DNA, the genetic mapping of chromosomes, cell cycle kinetics, cellular differentiation, tumour growth, control of biochemical systems, behaviour of physiological systems, radiation effects, etc.

Then at the population level we have models for demographic processes involving birth, death, and migration; population genetics; evolutionary phenomena; competition between species and predation; the population dynamics of communicable disease spread; statistical or 'functional' descriptions of noncommunicable disease incidence and prevalence; ecological systems and environmental factors, etc.

Such lists could be extended almost indefinitely. The essential aspect is, of course, the emphasis on improved scientific understanding. Most, but not all, of the subject areas mentioned above entail some accepted biological structure. Subsequent mathematical modelling is then aimed at providing a more precise, quantitative insight into the phenomena in question. From the point of view of useful applications in medicine and public health, our interest is not in mere mathematical refinements, but in the potentialities for improved practical control. This aspect is, however, covered by the other classifications below.

(*b*) **Operational and systems models**

The subjects of operational research and systems analysis will be dealt with *in extenso* in Chapters 7 and 8. For the moment, suffice it to say that their domain deals with man-made systems, rather than the naturally occurring systems which were the subject of the technical modelling just discussed in (*a*) above. The emphasis in operational research and systems analysis is accordingly on problems facing executive and administrative authorities, especial concern being shown for organizational patterns and processes, and the implications for management and decision-making. The quantitative models that have been developed thus arise out of a concerted, rational, scientific approach to such problems. Examples include applications to hospital ward design, operating theatre design, outpatient appointment systems, organization of ambulance services, management of blood banks, mass health screening, provision and organization of health services, health planning, etc.

We have already looked in more detail at the scope of the scientific method, as applied to health problems, in Section 2.2. We have also, moreover, emphasized the importance of the multisectoral approach in studying the interactions between several interrelated sectors, including health. The systems models referred to are therefore part of the quantitative basis of a scientific investigation of the relevant practical administrative control problems. There is no intrinsic logical difference from the technical models in (*a*) above, the distinction made being primarily one of convenience based on the broad subject matter.

It follows that there is no hard and fast line to be drawn between (*a*) and (*b*). A physiological system may easily be a complex system of interacting components, with feedbacks, time delays, etc., just like a health service. But the physiological system is much more easily seen as a topic for disinterested scientific enquiry, whereas the health service system immediately entails questions of practical efficiency, cost–benefit analysis, resource allocation, and public health control.

In addition, it often happens that the system analysis type of model can be viewed as a complex arrangement of interacting parts, where each part is itself a technical submodel. Thus the overall control of a communicable disease may require a systems model in which there are technical components dealing with demographic aspects, health manpower, the population dynamics of disease spread, the acquisition of immunity (naturally, by vaccination, or immunization), etc. Alternatively, a multisectoral model is by definition a systems model whose component parts are systems models dealing with the relevant sectors.

Systems models therefore cover quite a wide range of applications. At one extreme there are the Forrester system dynamics models (see Chapter 8), becoming progressively more ambitious through a series of applications to industry, cities, and the world. At the other extreme, there are tentative quantitative models designed to help a decision maker have a better feeling for the likely outcome of choosing one feasible strategy rather than another, and leaving him to weigh up the relative importance of the simplified mathematical picture, on the one hand, with the mass of human factors, moral, psychological, and social, on the other.

(*c*) **Information systems and communication models**

The concept of information systems is receiving widespread recognition at the present time, though opinions often differ considerably as to what, precisely, an information system is. There is probably a fair amount of agreement that mere collections of basic data, whether technical or administrative, do not in themselves constitute an information system. The key characteristic seems to be that in an information system appropriate data are collected, stored, retrieved, reported, or transmitted, for the express purpose of assisting managerial action. Or, to put it another way, an information system deals with the flow of information within an organization for the purpose of facilitating decision making.

Generally speaking, therefore, we are concerned here with what is primarily an administrative instrument. The kind of modelling involved is that exhibited by flow-charts showing the relationships and communication flows between various organizational entities. This often leads to the use of network analysis and PERT methods, and is principally concerned with scheduling, programming, and planning. Important though these techniques are, a detailed discussion of managerial and administrative techniques is outside the scope of the present book. We shall therefore not elaborate further. However, it is clear that many systems models, of the type referred to in (*b*) above, presuppose the existence of a relevant information system.

(*d*) **Policy analysis models**

Policy deals with the broad strategic problems facing top-level administrators and politicians. In the context of medicine and public health this essentially means health planning of a medium-term or long-range type, covering an administrative area of regional or national dimensions. All this is generally acknowledged to be a very difficult subject, although it is, of course, of vital importance to any community since it is concerned with their overall aims and aspirations.

In technical scientific work the primary objective is to understand how things happen in the natural world. This means that, however free and imaginative the spirit of scientific enquiry, investigation is inevitably bound by the constraints imposed by nature. At the same time, it is possible to harness scientific knowledge to the control of natural process and for the achievement of human goals.

There is a close analogy here with policy making and policy analysis. As indicated by Majone (1975), 'Policies are, in fact, tentative theories—about the nature of social processes and the working of social institutions—and policy analysis should accordingly develop as an organon for the critical evaluation of proposed solutions.' It follows that questions of feasibility investigated by some form of constraint analysis are of paramount importance, in order to determine the consequences and implications of the technical, economic, social, institutional, administrative, and political constraints imposed upon the policy makers.

The whole subject of policy analysis is still very much in its infancy, but it is evident that appropriate quantitative methods, backed by a scientific approach, are very much needed. As yet, the kind of modelling that can be used in this context is very simple. But in so far as policy analysis will have to be increasingly incisive in the future about the role of technical work, systems analysis, and information systems, ways will have to be found of integrating the corresponding modelling activities already discussed above in (*a*), (*b*) and (*c*).

A simple example of an international problem is given by Bailey (1975). The World Health Organization is increasingly involved in the coordination of health activities that occur of necessity at the world level. The provision of drugs and insecticides is just one area. We have already noted in Section 1.1 the urgency of dealing with the global malaria problem, some 350 million people still living in areas where the disease is endemic but where no organized antimalarial programs have yet been established, and in many areas the prevalence of malaria is rising. World-wide programmes of malaria control are estimated to require approximately 50,000 metric tons of DDT per year. On the other hand, WHO directly controls, through specific projects it is involved in, only about 100 tons per year—no more than one fifth of one percent of the total.

At the same time, supplies of DDT are more difficult to obtain, there are fewer manufacturers, and delivery times are increasing. It would still be possible for certain manufacturers between them to provide the supplies that the world needs, but they will only be able to do so if a guaranteed market can be assured and an appropriate delivery plan properly scheduled. Thus international cooperation between a variety of interests is essential, though not sufficient, for success.

This example shows how considerations of information about the extent of the disease, technical knowledge of the dynamics of spread, the economics of manufacture and delivery, etc., are all relevant to establishing and implementing a feasible and effective policy. Similar illustrations could easily be given for other communicable diseases, e.g. world control of recurrent influenza pandemics, or a range of parasitic diseases besides malaria of utmost consequence to the developing world. Noncommunicable diseases, such as cancer and cardiovascular breakdowns, are also of major importance and greatly in need of overall policy control.

The importance of the policy level cannot be overestimated, in spite of obvious past difficulties and failures that are all too easily pointed out. Major objectives can and must be established at the policy level. When clearer policy statements, backed both by a sympathetic appreciation of people's needs and a realistic acceptance of unavoidable constraints, begin to emerge, it will become easier to decide on the magnitude of the activities required at the lower hierarchical levels of systems modelling and technical modelling. Thus a proper understanding of the role of policy analysis and policy modelling can be crucial to an effective use of the more familiar technical and systems components with which this book is most concerned.

5.4 THE ART OF MODELLING

The preceding sections of this chapter have identified the essence of modelling, indicated important logical distinctions between different types of models, and discussed various hierarchical levels at which modelling may be applied. The whole modelling approach, with its associated techniques, is evidently of great power and is of almost universal application in its potentialities for assisting rational and scientific enquiry, as well as for exercising control over both natural and manmade processes.

At the same time, there is justifiable anxiety that the method will get out of hand. People may not properly understand what is going on, and reasonably fear that the tendency of some applied science to pursue an ultimately dangerous course may be intensified. Alternatively, and this is particularly true in the health field, it may be thought that esoteric modelling involving expensive computerization will be both intellectually confusing and wasteful of resources. What, then, should be our overall attitude to modelling? How do we ensure that we know, at least in outline or in principle, what is proposed or what is being done? How do we evaluate the work of supposed experts in the field? How do we maintain an expeditious balance between abstract theory and practical utility? And so on.

These concerns render essential some discussion of the art, as opposed to the purely technical use, of modelling. We must be careful not to fall into the trap of the 'medicine-is-an-art' defence against rational enquiry mentioned at the beginning of Section 3.3. What is proposed here is an explicit discussion of the attendant difficulties with a view to resolving contradictions and misunderstandings. Only in this way can we expect a healthy development of modelling activities in a climate which is both sympathetic and constructively critical.

In principle, there ought to be little difficulty. We have already stressed the importance of testing the adequacy of hypothetical mathematical models in relation to actual data. This testing of hypotheses is an integral part of the scientific method. Any worthwhile scientific hypothesis has at least some quantitative aspects, but we are more concerned in this book with those formulations that are sufficiently detailed to merit the title of 'mathematical model'. The logical status of such a model is, nevertheless, precisely the same as any other less highly quantified hypothesis. That is, it purports to be a partial account of certain aspects of reality, and its validity depends entirely on the accuracy of this account.

Because of the extreme complexity of most biological and social systems, the simpler and more manageable mathematical models are easily recognized as gross oversimplifications. And, worse than this, it may well happen that enthusiastic mathematicians will embark on the extensive theoretical development of models known in advance to be inappropriate simply because the work, to them at any rate, is easy. Alternatively, an inherently plausible model is seized upon and subjected to mathematical analysis but fails to satisfy the canons of the scientific method because there seems to be no way of testing the conclusions. A good deal of current work is open to objections of this sort, which if sustained would be highly damaging. Biologists, clinicians, and public health authorities are rightly suspicious of any work that seems to be merely the spinning of mathematical abstractions and not to be capable of incorporation within a continuous process of scientific advance.

Our claim is that, correctly understood and applied, a mathematical model has the same logical status as any other scientific hypothesis, and must therefore be manipulated and justified according to the same criteria. Precisely because the model has greater clarity, the attendant difficulties are more readily perceived. Part of the answer to these problems is to find better ways of handling models that are sufficiently elaborate to be of adequate realism, as in the computerized simulations of complicated stochastic processes. This is not, however, the whole story. At the end of Section 4.5 on stochastic processes brief mention was made of the importance of simple models, having only a few realistic features, in providing general insight, while more elaborate models would be required for actual statistical fitting to data. This conflict of the requirements for insight and reality appears to be fundamental in mathematical biology, and it is important to appreciate the full significance of the difficulties.

Thus, at an elementary level, we may succeed in using a crude deterministic exponential law of population growth for demographic purposes, such as making rough predictions of the future population size when numbers are large. More accurate results can be obtained by stratifying the population into age and sex groups each with their own individual rates of increase or decrease. Now suppose that we want to study smaller groups in which statistical fluctuations are very much in evidence. As seen in Section 4.5, there are appreciable complications in dealing with even the simplest stochastic birth process, let alone the difficulties introduced by death, migration, spatial differentiation, etc. Moreover, most mathematically tractable models seem to make unwarrantable assumptions, such as postulation of exponentially distributed lifetimes. More realistic distributions make the processes considerably harder to formulate and handle. Nevertheless, the important point is that even the oversimplified models have their uses, and are scientifically valuable up to a certain point.

Having started, for example, with a deterministic model, we then want to know the consequences of introducing probability elements via the appropriate frequency distributions. Are these likely to have large or small effects? What general properties would a stochastic model be expected to have as compared with a deterministic one? And so on. These questions can be validly answered in a general way by investigating some simple stochastic processes, and examining such quantities as means, variances, chances of extinction, distributions of total population size, etc. This, we hope, will provide a broad measure of insight. We may decide that certain aspects must be included and that others can be neglected. In studying mutation in bacteria, for instance, it is essential to adopt a stochastic treatment of the small groups of mutants since these embody large amounts of statistical variation, but in many cases the main population of normal individuals is quite large enough for a simple deterministic treatment to be an adequate first approximation. Again, the simplest stochastic investigation of the phenomenon of recurrent epidemics shows that this kind of mathematical description is capable of accounting broadly for the important observed feature of repeated, undamped outbreaks, while the deterministic theory involves a series of damped oscillations and is strikingly unsuccessful. If we now wanted to develop more elaborate and realistic models of bacterial mutation or recurrent epidemics, the information gained from the preliminary, simplified models would be most valuable. Ultimately, of course, the whole line of research stands or falls by the competence of the models developed to explain and predict real observations.

One of the great advantages of a well-organized mathematical model is that it entails a considerable amount of precisely described structure. On the one hand, this lends itself to practical verification by suitable physical, chemical or biological experiments. On the other hand, mathematical analysis usually adds to our theoretical knowledge about the structure. Whenever basic equations can be solved in mathematical terms, we automatically solve all possible problems obtainable by substituting different values for the parameters involved. Further, the algebraic form of the solutions may in itself lead to new knowledge in the form of mathematical theorems. The threshold theorem in the theory of epidemics is a case in point. Another example is the conditions for equilibrium in a population where advantageous and disadvantageous mutants are genetically linked.

Wherever possible one utilizes a purely mathematical development of the model, as this affords the greatest possible analytical power. Unfortunately, the basic equations often prove to be mathematically intractable, and we have to fall back on numerical solutions. The subject of numerical analysis, which is discussed further in Section 6.4, is full of pitfalls for the unwary. Nevertheless, given adequate care, a range of computed solutions will often supply a considerable measure of useful information about the properties of a model. As models become more complex in order to achieve greater realism, so the number of instances diminishes in which neat, elegant, explicit solutions can be found. Some form of numerical investigation is becoming increasingly obligatory. Hence the central importance nowadays of high-speed electronic computing facilities.

In some cases the difficulties are even greater. We may be able to formulate the equations of motion for some complicated biological process, but find that the resultant differential equations, perhaps multidimensional partial differential equations, are not only mathematically intractable but also are unsuitable for solution by existing numerical analytic methods. A simulation study, of the type discussed in Section 4.5, may then be the most profitable method of investigation. Simulation studies share the disadvantage of standard numerical analysis in not providing algebraic descriptions of the implications of the model. But they also have the countervailing advantages of simplicity in operation, flexibility in accommodating minor changes in the model, and an avoidance of many tricky numerical niceties by relying on the statistical properties of a sufficiently large number of repetitions.

Another point worth mentioning in connexion with the use of simulation is that, although in principle it does no more than could be ascertained by purely mathematical derivation, in practice it brings one much closer to the conditions of real experimentation. This is increasingly important in the study of complex biological phenomena, and also has extreme relevance to the complex processes investigated by operational research and systems analysis (see Chapters 7 and 8). In many practical operational situations we wish to implement some fairly optimal scheme of action, but in the nature of things are unable to carry out real experiments in any extensive way. Quasi-experiments, however, can be performed on simulated systems which have been made sufficiently elaborate to achieve a high degree of realism.

In ordinary technical level scientific work, whether in biology or medicine, any excessive zeal for mathematical modelling and computerized investigations will be automatically curbed sooner or later by a failure of the hypotheses or theories to account for the real world. However, in a strictly scientific context, where a continuous succession of experimental undertakings is the accepted standard, serious harm is unlikely to result.

Greater dangers lie in the analysis of large systems for decision making and planning purposes. Administrators and planners want to base their decisions on what would happen if certain changes were made in the existing system. Inevitably, as pointed out in Section 5.2 under the heading of 'Deductive and inductive models,' heavy reliance is placed on deductive methods. But, as pointed out, much more needs to be done by way of inductive validation of the theoretical models assumed *for the time being* to represent the best knowledge available. It is *absolutely vital* that decision makers who rely on modelling aids be aware of this aspect.

If the commonsense view of the scientific approach, mathematical methods, and modelling techniques adopted in this book were generally accepted, there would be little difficulty. Questions would be continually asked in any situation as to whether the models used were sufficiently simple for practical applications while still realistically representing the salient features of the real world; whether all main unknown parameters could be estimated from data, or whether a considerable amount of inspired guesswork was involved; whether the models could be fitted to actual observations, and whether sufficiently accurate forecasts could be made of the future behaviour of the system.

These dangers are increasingly compounded by the practice of constructing computerized methods of handling complicated systems models in an easy and painless manner. Because a large number of relevant facts have been fed into a computer, some adminstrators may believe that the resultant printouts consist of reliable conclusions. In such cases it is forgotten that the conclusions may be highly dependent on the assumptions which also have to be fed into the computer, and which constitute the underlying model. If these underlying assumptions were explicitly reviewed by the decision-making user, he might easily fail to understand them, in which case he would have no idea whether the conclusions were reliable, or he might understand them well enough to disagree with them. Either way, the situation would be highly unsatisfactory.

The use of 'interactive' computer methods, while extremely attractive and potentially very powerful, encourages the development of the foregoing dangers. The user operates directly a simple keyboard rather like a typewriter, and can easily write in his own questions or respond to questions printed out by the computer. With careful advance programming of the computer, and the use of specially prepared computer software packages, a dialogue situation is set up between the computer and the operator. With a little practice, this can be a very stimulating experience. Important answers to serious questions appear to result from the ensuing 'conversation' taking place in ordinary language, but possibly using such technical terms as the user has himself defined.

While the potential power of such methods should not be decried, a highly illusory confidence in the computer can be built up. This is perhaps more a question of the abuse of computers, to be discussed in Chapter 6. It is, however, specially mentioned here because of the circumstance of various underlying mathematical models being programmed into the computer by the software employed, and not being readily accessible to the user for critical consideration.

Similar objections hold in relation to the use of interactive or conversational modes for teaching purposes. Models may be implicitly assumed that students cannot understand or are not aware of, yet they can learn to extract from the computer the semblance of answers to questions they have posed.

Examples of difficulties and pitfalls in the use of modelling could be extended indefinitely. There is no scientific answer to this problem, since it is really the use of the scientific method itself that we are talking about. Hence the reference to the 'art' of modelling. Perhaps the best guide to practical success is a constant logical and philosophical critique of ongoing modelling activities, coupled with plenty of commonsense plus 'eternal vigilance'. In accordance with the well-known principle of William of Occam's razor, *entia non sunt multiplicanda praeter necessitatem*. In other words, don't make things more complicated than they have to be!

CHAPTER 6

The Use and Abuse of Computers

6.1 GENERAL PRINCIPLES

Except for the gifted few, mankind has since the earliest times regarded arithmetic calculation as an immense source of drudgery, to be avoided at all costs and by any means available. The abacus, essentially a primitive hand-operated digital computer, has a long history, and seems to have been invented in several different parts of the world independently and is still used with good effect both for financial and mathematical calculations in Eastern Europe and the Far East. With the invention of logarithms near the end of the sixteenth century, the slide-rule became a natural and inevitable mechanical development. The first one appeared in the 1620s. Far more subtle calculations were then possible, though the analogue nature of the device (using physical distances to represent numbers) severely limited the accuracy for many types of work. Nevertheless, the slide-rule is still a standard piece of equipment for mathematicians, scientists and engineers the world over. Again, one thinks of more specialized aids such as the accountant's ready-reckoner, the grocer's scales or the visual display of a modern petrol pump. In all of these devices pre-calculated tables or scales are used to effect the more or less instantaneous multiplication of a given sum of money simply by entering the correct table or observing the appropriate scale at an easily recognized point.

Such gadgets are now commonplace, and are widely accepted and understood as straightforward tricks for avoiding much simple though tedious arithmetic. At a slightly higher level of complexity there are office adding machines and cash registers which will easily perform a long series of additions and subtractions and will keep a printed record of operations performed, together with various totals and subtotals. An extensive range of more advanced accounting machinery has been developed from these relatively simple pieces of equipment. Although the operators may feel that they do not fully understand how such apparatus works, they usually have no difficulty in appreciating that they are merely using a convenient mechanical device for doing more efficiently what they themselves could do with pencil and paper, given sufficient time, an absence of fatigue and error-free calculation.

Similar considerations also apply to the more elaborate desk calculating machines. To begin with, these were basically mechanical digital devices for performing the fundamental processes of arithmetic, namely, addition, subtraction, multiplication, and division, together with a number of ancillary controls for accumulating results and carrying out various short-cut procedures. The cheaper machines were normally hand-operated and were extremely useful for relatively small jobs and for teaching purposes. For very lengthy calculations, where greater speed and flexibility are required, larger and more expensive machines were necessary. These were usually driven by an electric motor, though still essentially mechanical in mode of operation. Desk calculating machines of this general type were in use for a great many years, and until about 20 years ago were the chief means of carrying out practically all scientific numerical calculations of any appreciable degree of complexity. Nowadays, mechanical desk calculators are being rapidly replaced by faster, cheaper, and more versatile equipment based on modern electronic technology.

The advantages of a desk calculating machine over ordinary pencil-and-paper methods are enormous and well known. Numbers can be introduced into the machine quickly and accurately by setting levers or pressing buttons; all actual arithmetic is performed internally by the machine itself; and with careful planning of the sequence of calculations a great deal of the writing down of intermediate results can be eliminated. Mistakes can of course occur, though less easily than with pencil-and-paper work, and it is standard practice to incorporate a number of checking procedures into the sequence of calculations.

The availability of a desk calculator naturally extends the horizon of feasible computation, but there soon comes a time when a natural limit to overall speed of operation is reached. It is easily verified that the limiting factor in performing a long series of calculations, say accumulating a large number of products, is not so much the speed of the machine itself as the time taken to put numbers into it, to read and transcribe results, and to decide what to do next. The overall time taken to perform a complicated task might therefore improve but little, even if the machine executed the basic arithmetic infinitely quickly. Again, as the complexity of the work increases so the opportunities for fatigue and error increase. Moreover, if a particular kind of calculation is to be repeated in general form a large number of times, a correct sequence of operations in one calculation is no guarantee that a mistake will not be made in subsequent repetition. Another difficulty is that it is usually advantageous from the point of view of time and accuracy for the main work of a lengthy calculation to be undertaken by an experienced machine operator, although this means that the originator of the calculation will have to expend much effort in preparing the work and explaining to the operator what is required. And if the same mathematical method is to be employed again later with different data, these explanations may have to be repeated, perhaps to a different operator.

While desk calculating machines represent a very great advance over previous methods of computation, the shortcomings are sufficiently evident for there to have been a strong motive for developing a rather different kind of machine. An appropriate form of electronic operation will obviously provide a vastly quicker method of carrying out the basic arithmetic than is possible by purely mechanical means, but it is not feasible to take full advantage of this unless human intervention can be reduced to a minimum. This means that not only must the slow recording, reading, and transferring of numbers by humans be largely eliminated, but the whole plan of executing a calculation must somehow be arranged so as to avoid appealing to a series of human decisions while work is in progress. It is in the solution of these latter problems, rather than in the mere speed of internal operation, that the real revolution effected by the modern automatic electronic computer consists.

These machines began to be developed towards the end of World War II. The first to incorporate the major design features of modern equipment were EDSAC, which became operational at Cambridge University, England, in 1949, and SEAC which was completed at the National Bureau of Standards in the U.S.A. in 1950. The early machines all used thermionic valves, but these were later superseded by transistors which permitted smaller computer sizes and far more reliable operation. Big advances have taken place over the years in the microminiaturization of equipment. This has led to the availability of electronic computers on the scale of desk equipment, and speeds of basic operations measured in thousand-millionths of a second (nanoseconds). For detailed accounts of what can be achieved, and how precisely to make use of available equipment, the reader should consult some of the many excellent textbooks that now exist. We shall content ourselves for the most part with discussing the main principles in so far as they affect the central themes of this book. The special consequences of the recent appearance of pocket-sized electronic computers are discussed separately in Section 6.6.

Let us examine the salient features of a modern automatic electronic computer. In the first place it is important to realize that basically it still performs the same kind of arithmetic operations as the desk calculator, and therefore in essence does no more than is in principle possible for a person working only with paper and pencil. The differences lie in an enormous improvement in technical efficiency on the one hand, and in the logical control of the sequence of calculations on the other. What the computer is ultimately doing should therefore be no more difficult to appreciate than arithmetic tasks carried out in a more mundane fashion. Whether one regards the activities of an electronic computer as showing signs of intelligence or as being closely analogous to cerebral cortical behaviour depends largely on definitions. The fundamental point is that however elaborate and complicated the computer may be it is essentially a highly developed tool for certain specific purposes, and as such is to be regarded in much the same light as any other piece of complicated scientific apparatus like an electron microscope or a linear accelerator. Moreover, the computer will only do, at best, what it is told to do.

A standard mechanical desk calculating machine holds within itself the pair of numbers one has fed into it for some specific arithmetical operation, e.g. addition, and may still retain these numbers after it has carried out the operation. It may also hold one or two other numbers such as the accumulated sum of repeated additions or multiplications, or the accumulated total of multipliers. But the total storage capacity rarely amounts to more than five or six separate numbers, and even then these are usually severely restricted in size. The electronic computer, on the other hand, is typically well able to store several thousand numbers to a good degree of accuracy in a fast-access form using some appropriate electromagnetic device, and to store several million numbers in a relatively slow-access form (slow by electronic standards) on magnetic discs or tape. This provides great opportunities for eliminating human intervention in handling intermediate results, and also permits the automatic processing of very large quantities of numerical data. Once the latter have been recorded on punched cards or paper tape, or on magnetic discs or tape, they can be used repeatedly without human mental or visual effort simply by operating the relevant peripheral reading device on the computer.

Next, and this is really the heart of the matter, it has been possible to develop ways of planning a whole sequence of calculations so that a *program* of instructions is fed into the computer and stored along with the relevant data. These instructions are written in a suitable numerical shorthand and primarily relate to performing basic arithmetical operations on pairs of numbers in specified parts of the computer's store. Once the program and the data have been fed into the machine all further operation is then purely internal, and can be effected at speeds depending on electomagnetic design until the final results are punched out on cards or paper tape or fed directly to a teleprinter or other display device. Much careful thought may be required in writing the program for any given sequence of calculations, but when the program has been developed and properly tested, it can be used repeatedly without further checking. The saving of time and labour is clearly considerable. Thus an adequately tested and checked program might be used thousands of times by different people working on different computers. Again, a no less important aspect is that a given program may involve an instruction to carry out a particular task, like solving a certain set of equations, several hundred or thousand times using different data. Part of the program must deal with the solution of a single set of such equations. When this part is correctly written it can be used as often as we please within the program with complete confidence that the correct method will be followed each time. This overcomes one of the outstanding disadvantages of desk machines already mentioned above, namely that the correct performance of a desk calculation on one occasion is no guarantee against inaccuracy in repetitions on subsequent occasions.

With a desk machine one is of course closer to the calculations in progress at all stages. If something strange and unforeseen occurs, steps can be taken to investigate it immediately, and serious errors are often avoided in this way. This circumstance is frequently used as an objection to computers on the grounds that these provide so many greatly increased opportunities for unobserved errors at intermediate stages that the final results must be highly questionable. Fortunately, this difficulty can be largely eliminated by good programming. In ordinary desk calculations there are, as we have seen, very great opportunities for repeated error. Only the grosser mistakes are in fact immediately apparent, and we can only guard against the accumulation of small errors by careful checking at all stages. These checks are an essential part of the whole sequence of calculations, and must also form an integral part of a program of instructions for an electronic computer. Indeed, because of the speed of operation of electronic computation it is possible to incorporate many more arithmetic checks than are normally regarded as feasible in desk calculation. A well-designed program for an electronic computer is accordingly capable of yielding results that are far more reliable, not less.

We have of course been considering above the possibility of error arising from faulty control of the computer rather than from purely electrical faults. Some check against the latter occurring without our knowing it can be made by, for example, designing the computer in such a way that it will automatically stop if an error occurs in the so-called 'parity' of a number. In most machines numbers are actually held in binary notation, as a sequence of zeros and ones. An additional checking, or parity, digit can be added which is zero or one according to whether the total of ones in each stored number is odd or even. At certain times, e.g. when reading a number from the store, the parity digit can be checked against the number itself. Such devices are not completely foolproof, and compensating errors can occur. But in practice they go a long way towards drawing attention to electrically caused errors.

So much then for some of the main ideas involved in the use of automatic electronic computers. The widespread availability of such apparatus has a far-reaching effect on the kind of calculations that are now feasible, the scale of data-processing that can be contemplated, and indeed on the general formulation of research projects and the choice of techniques to carry them out. This aspect is taken up again in Sections 6.5 and 6.6. In the meantime we shall take a closer look at some of the main tasks for which it is worth using a computer, and the methods that can be employed for enabling research workers to take advantage of the facilities available.

For an excellent introduction to the whole field of computing science the book by Hollingdale and Tootill (1975) is to be recommended.

6.2 DATA PROCESSING

We have already indicated in Section 6.1 that besides an enormous capacity for high-speed arithmetic (with the implications for numerical analysis to be elaborated in Section 6.4) the electronic computer also exhibits the major facility of storing huge quantities of data. This feature has led to the development of a broad area of interest, loosely referred to as *data processing*, which is to a large extent distinct from numerical analysis, though the two disciplines certainly interact and complement one another.

The compilation of extensive records has long been a common practice in most scientific subjects as well as in law, commerce, industry, government, etc. Records may be used administratively to establish the rules and precedents for social behaviour or action, or they may be used to suggest, formulate, and test scientific hypotheses. In some cases it is possible and convenient to 'reduce' large quantities of numerical data by using an appropriate statistical method, as discussed in Section 4.1. Once the relevant statistical indices have been computed they are available for further numerical development without the necessity to consult the original material again in detail. Frequently, however, the basic records must be retained in something like their full form for testing alternative methods of investigation and because individual items may be of crucial importance.

As is quite familiar, the simplest types of record-keeping involve collections of original documents, including anecdotal information as well as quantitative measurements. If there are a great many documents some kind of filing system must be used, together with a card-index constructed to give rapid entry to all material classified under any given heading. The vast majority of medical records, both in hospital and private practices, still exist in this form. While it may be relatively easy to find the records for any particular patient, there are usually well-nigh insuperable difficulties in presenting summary data in the shape of tables showing the frequencies of various items or the association between different aspects of the material, unless special steps are taken to facilitate such analysis. A small number of factors can be studied more easily by keeping a running check on the relevant items as the records are amassed. For example, the patients entering a given inpatient department might be recorded in a special book showing age, sex, diagnosis, length of stay, clinician in charge, etc., and various cross-tabulations might be brought up to date on discharged patients every week or month. To carry out this type of investigation by ordinary clerical methods is laborious, time-consuming, often inaccurate, and sets a fairly low limit to the volume of material that can be handled and the number of techniques of analysis that can be tried.

Considerable improvement can be made by recording the relevant data on special cards in which the coded classifications are registered by suitably punched holes. These holes may be punched round the edges of the cards, as in the Cope-Chat and McBee systems, so that sorting can be carried out manually on an office desk with the aid of a knitting needle or similar instrument. Thus a hole may represent the absence of a certain factor. If the factor is present the hole is converted to an open slot or notch. By pushing the knitting needle through the appropriate position for a stack of cards and lifting, we leave behind all cards with notches, corresponding to a group of individuals for all of whom the factor is present. Suitable combinations of holes can be adapted to provide numerical or alphabetical classifications.

The next stage in the process is to utilize the whole surface of the card, as in a typical IBM card with 80 columns each of 12 positions. Considerably greater quantities of information can now be recorded, and such cards can be sorted and counted automatically using an electronic sensing device which can tell if any given position has a hole punched in it. These punched-card records are a convenient addition to the original medical documents for both administrative and research purposes. The work is still relatively slow and cumbersome, though a great improvement on desk-sorting methods. Routine administrative returns can be compiled very conveniently provided that they are not required too frequently. Annual or quarterly figures are in general perfectly realizable, but daily results are in most cases not feasible. Preliminary research investigations can be undertaken without too much delay, but if extensive cross-tabulations are required, especially when elaborate statistical calculations are to be made, the whole process is liable to become unmanageable.

Tremendously improved methods of handling such records are now provided by typical electronic computers. Data may be transferred from original records to punched cards in the first instance, as a matter of convenience. The cards are electronically read and the information stored on magnetic discs or tape in a very compact and accessible form. All subsequent sorting and analysis are then performed by the computer at very high speed. Although stored data have to be read, when required for analysis, from disc or tape, this can be done extremely rapidly. Moreover, large numbers of summary tables or other forms of analysis can be carried out during a single search of the stored material. This may be compared with the laboriously repeated sorting of original punched cards required by the mechanical methods of classification described in the previous two paragraphs. If only a few hundred records are to be manipulated, it may be quite sufficient to sort these out on one's office desk, using either the original documents or punched-card versions. But with anything appreciably larger, involving a few thousand records, say, it will nearly always be worth the extra initial effort to transfer the material to some kind of electromagnetic storage and to construct appropriate computer programs for handling the data in this form. This entails a great improvement in speed, accuracy, and flexibility. Speed hardly needs special comment. Increased accuracy results because of the reduction of human intervention and the possibility of building in many more cross-checks and consistency tests. Greater flexibility means that a far greater range of analyses can be undertaken, and that the purely sorting or data-processing aspects can be used as preliminaries to full-scale scientific computation of the type discussed in the previous paragraph.

In addition to the processing of data that have already been compressed and compiled into conventional handwritten or typed records, there is the problem of dealing with the extensive raw material such as occurs with electrocardiograms, electroencephalograms, and other similar physiological records arising in man or in animals for experimental or diagnostic purposes. There are a number of different kinds of possible analysis depending on the requirement of individual problems. All are essentially concerned with improving the signal-to-noise ratio so as to bring into greater prominence significant underlying waveform patterns. Thus the use of a response-averaging procedure will give a clearer picture of a repeated waveform as a complete entity. This may be of value if specific patterns are known to have diagnostic significance. Alternatively, we may decide to try to identify basic oscillatory components as exhibited by a Fourier type of frequency spectrum analysis. Again, because of the time-series nature of the basic biological process an autocorrelation approach may be preferred. All these techniques can now be handled by on-line procedures, in which the observed electrical potentials are fed directly to the appropriate kind of computer. For some purposes analogue machinery is quite convenient and is frequently used as a first stage. But for the application of a range of sophisticated mathematical methods conversion of the data to digital form is essential.

As a further extension of on-line working we can consider the application of the foregoing methods to automatic patient monitoring. Systems have been developed to enable a continuous display of several types of clinical observation to be obtained, such as temperature, heart rate, blood pressure, respiration, etc. In the first instance, of course, the problem is to find satisfactory methods of continuously measuring the relevant activities, of compiling permanent records as well as presenting suitable visual displays, of incorporating 'alert' and 'alarm' indicators, and so on, without necessarily using any elaborate computing techniques. The latter might, however, be very usefully included in cases where, for example, some type of waveform analysis was desirable for electrocardiograms or electroencephalograms. In general, monitoring systems are conceived as providing a very rapid indication of possible deterioration in the state of critically ill patients. Suitable action must be taken by nursing or medical staff whenever alert or alarm situations occur. But it may well be possible to arrange for certain kinds of clinical action to be taken automatically in cases when speed is essential.

Another type of data processing is afforded by the field of pattern recognition, as for example in the automatic classification of chromosomes. Electronic scanning methods are employed to assess photographic data in such a way that the actual shapes can be converted to digital form and subjected to appropriate computational analyses.

For an introductory discussion of various developments in the area of data acquisition and processing the reader might like to consult the published proceedings of the series of Rochester Conferences edited by Kurt Enslein (1962 and later).

6.3 COMPUTER PROGRAMMING

When decisions have been made about the general type of problem we are dealing with and the form of analysis that appears to be appropriate, whether in the field of data processing or numerical analysis or some combination of the two, we are faced with the specific task of instructing the computer to carry out our chosen sequence of calculations. As indicated in Section 6.1 This is done via a suitably designed program of orders. An understanding of computer programming is accordingly the central activity in the utilization of automatic computers. It is customary to employ specially trained programmers to write new programs or adapt existing ones, in the same way that it is advantageous to have expert operators when we want to make extensive use of desk machinery. In both cases, however, it is usually important for the originator of the problem to know something about the type of machine available, as this may critically influence the choice of analysis. This point will be pursued further in Section 6.5. For the moment we are merely concerned to discuss some of the salient features of computer programming for the benefit of those who may be unfamiliar with the implications.

To begin with, we observe that although most electronic computers represent numbers internally in binary form, i.e. in an appropriate sequence of 0's and 1's, for programming purposes we can write our numbers, whether in data or in formulae, in the usual familiar decimal notation. When these numbers, recorded on cards or tape, are fed into the machine, they are automatically translated into the relevant binary code. Again, the various locations or registers in the machine's store may hold either numbers or instructions to carry out certain arithmetic operations. The instructions look exactly like numbers (the machine has to be told which interpretation is correct) and are also in binary form. For convenience, therefore, a special code is adopted so that we can easily tell the machine what arithmetic operations to carry out. The arithmetic unit in a computer normally contains an accumulator register just like a desk machine. So, to add two numbers which are already stored somewhere, we first instruct the machine to transfer one of them to the accumulator, then add the second number to the first, and finally transfer the sum back to a convenient place in the store.

Consider, for example, the following typical *machine code* instructions. The individual locations in the store are all numbered, and each may contain a number or an instruction. Suppose that two numbers, x and y, are stored in specific numerical form in locations 251 and 252. We want to form the sum $x + y$ and put it into location 253, leaving x and y intact in their original positions. Let the instruction code for copying a number into the accumulator be '30' followed by the number or address of the location where the number is stored. The first order is therefore '30 251', which copies x into the accumulator. We next add y to what is already in the accumulator. If addition is indicated by the code '04', then the complete second order is '04 252'. Lastly we want to transfer the sum $x+y$ now in the accumulator back to location 253. Let this operation follow from instruction '20'. We thus have '20 253'. The sequence of machine code instructions are written on a program sheet as

30 251
04 252
20 253

and we end up with x in 251 and y in 252, as at the beginning, and the new quantity $x+y$ in 253. This is of course only a very simple example, but it shows essentially the sort of way in which one communicates instructions to the computer. It should be mentioned here that the instructions on the program sheet must in practice be stored somewhere in the computer's storage unit. A so-called control unit then takes them out, one by one, interprets and executes them, in general in the order in which they are stored.

Although there are only four basic arithmetical operations, addition, subtraction, multiplication, and division, it is convenient to specify a wider range of combined operations. Thus we may want to transfer a number to the accumulator and clear the location from which it was drawn; or we may want to add a number to what is already in the accumulator and transfer the result to the location from which the number was taken. A typical machine instruction code has some thirty or forty alternative combinations of this kind, and provides a simple, flexible, and easily memorized code (at least if one uses it constantly).

Since all numerical calculations are built up of basic components of this sort, it is clear that we can compose a machine code program to deal with any ordinary sequence of arithmetical operations. All the things we might do with a desk machine, including reading in numbers, transferring or copying them, have machine code equivalents. Of course, for lengthy calculations the writing of machine code programs is a time-consuming and tedious task, though if well done it does provide an efficient set of instructions—efficient in the sense of utilizing to the full the computer's potentialities for rapid calculation. Programmers with a lot of experience can become very fluent in writing good machine code programs. But for users who do not have the time, or perhaps ability, to acquire this skill, some simpler means of communicating with the machine is required. For this purpose, various kinds of programming languages, or *autocodes*, have been developed, in which instructions are written in familiar verbal or algebraic form and are then translated by the machine itself into the equivalent machine code versions.

Thus an autocode version of the foregoing machine code example could be written quite simply in terms of X, Y and a third number which we could call Z as

$$Z=X+Y.$$

Strictly speaking this means that there are three machine locations called X, Y, and Z; and the expression $Z=X+Y$ is not an equation but an instruction meaning 'replace the contents of location Z with the sum of the contents of locations X and Y'. If we wanted to store the sum in the location originally occupied by X, we could write the instruction as $X = X + Y$, which indicates more clearly that we are working with directives that are algebraic in form though they are not really algebra in the true mathematical sense. There is, however, such a close correspondence between this symbolic coding and true algebra that skills acquired in handling the latter are easily and rapidly transferred to the former. This is why autocodes are so popular.

In point of fact, machine-code programming has largely disappeared so far as big modern computers are concerned because of the highly sophisticated forms of engineering construction now adopted. Instead, the lowest level programming is done by programming specialists in so-called *assembly language*. This is a very simple type of autocode designed to make maximum use of all the specialized facilities of any given machine, and therefore its usefulness is liable to be confined to the machine for which it has been developed. Some machine-code programming has, however, recently reappeared in the use of certain programmable pocket computers (see Section 6.6).

It is evident at this stage that there is a very clear and fundamental distinction between the computer itself, seen as a complex system of electronic gadgetry, referred to as *hardware*, and the mechanism of communication constituted by an appropriate collection of computer programs, referred to as *software*, by means of which a user instructs and controls the computer.

Let us now consider, by way of illustration, one of the most widely used autocode programming languages, BASIC, which is frequently used for the control of small computers such as programmable desk machines. Suppose we want the computer to read two numbers x and y, in that order, from an input tape, and print them out with the additional quantity $\sqrt{(x^2 + y^2)}$. A complete program for doing this could be:

```
READ  X, Y
PRINT  X, Y;  SQR(X↑ 2 + Y↑ 2)
```

There is no need to discuss this in detail, but we note in passing the special device '↑' used to indicate a power. Special conventions like this are adopted for many mathematical symbols so that the typing or printing involved can be achieved by writing on a single line without recourse to suffixes or superscripts.

Simple programs like this are easily understood, and by suitably putting together elementary forms of computer instructions, quite sophisticated programs can be built up to cope with highly complex calculations. It must of course be emphasized that the 'rules of the game' have to be observed rather strictly. The grammar and syntax of the programming language must be accurately observed, otherwise the computer may misinterpret the programmer's intentions. If some purely logical error is committed the computer may print out a warning message and wait for a correction; otherwise it will go happily on computing the correct answer to the wrong problem!

This is because the symbolic instructions in the autocode are translated automatically into machine-code instructions by a special program, or *compiler*, that has been designed for this purpose. The compiler can thus deal only with syntactically correct expressions, but will itself be programmed to generate error messages when formally incorrect expressions occur. Therefore, provided the autocode instructions are correct in *form* the compiler will mechanically translate them into machine code. It is essential to realize that the compiler has no way of telling whether the instructions have meaning or are senseless—this is entirely the responsibility of those who are using the computer.

One of the most powerful aspects of automatic programming is the way in which instructions for a long series of repetitive operations can be described very simply and compactly. For instance, suppose we have a data tape with N numbers $x_1, x_2, \ldots, x_N$, recorded on it, and we want to know the average value of all the x. A convenient way to do this is for the computer to read the data values in sequence, forming a cumulative sum as it goes along. If the final value of this sum is S, we want S/N printed out. Suitable instructions might go as follows:

```
S = 0
FOR I = 1 TO N
READ X
S = S + X
NEXT I
PRINT S/N.
```

The part of the program between FOR and NEXT is carried out repeatedly for $I = 1, 2, 3, \ldots, N$. Each time the instruction READ X is reached a new number is read off the tape, and in the following line is added to whatever total is already in S. To be sure of starting with $S = x_1$ in the first operation with $I = 1$, we put $S = 0$ before embarking on the repetitive cycle or *loop*. Otherwise, we should be adding x_1 to whatever number had been left in location S by some previous calculation. Sometimes all locations can be cleared automatically before starting, but this may not be possible for intermediate calculations. In any case, it is best to play safe with $S = 0$ at the beginning. When the loop indicated is completed, the computer goes on to process the next part of the program, which in this case is just to print out the average S/N.

In practice, each line of a BASIC program must be preceded by a line number. These must be written in ascending order for internal 'book-keeping' purposes. The line numbers have been omitted from the above examples for simplicity, but in an actual program they introduce a large element of flexibility since one can move freely from any one part of a large program to any other by introducing an appropriate shift instruction. Thus, if at some point it is desirable to go next directly to the instruction in line 57, one simply writes 'GOTO 57'.

There are of course many types of autocodes, which in the past were often fairly specific for a particular kind of computer or closely related range of computers. Attempts have been made therefore to construct programming languages that are more universal, and can be used with quite different machines. This means that programs developed on one machine are available for controlling calculations on any other machine that can translate the universal language into its own special machine code. In Europe the programming language ALGOL (from ALGOrithmic Language) is widely used. In the first place, this is a more powerful and elegant language than any autocode in which scientists and programmers can write their computational procedures. Secondly, a great many different types of automatic computers now have ALGOL translators (or *compilers)*, though there are often small differences between machines as regards individual limitations. Many machines cannot handle all the facilities of the most general form of ALGOL, and the input and output arrangements usually differ from one machine to another. Nevertheless, a very valuable degree of universality and interchangeability has been achieved.

Another very important language extensively employed in the United States, as well as in Europe and elsewhere, is the IBM language FORTRAN (from FORmula TRANslator). There is considerable argument between the proponents of FORTRAN and ALGOL about their relative merits. It is not the purpose of this book to debate this kind of question, though it would be quite fair to state that both languages are highly suitable for scientific computing and have been widely used with great success. The situation *vis-à-vis* data processing is more complicated, and the choice of language depends on how much data handling and how much numerical analysis are required. For the type of applications commonly required in business, COBOL (from COmmon Business Oriented Language) is a very suitable language, being composed largely of words and phrases occurring in ordinary English usage, but including some elementary arithmetical operations as well. There are also several other popular programming languages, including PL/1, APL, etc.

The task of actually *coding* a computational procedure in an autocode or other symbolic language like ALGOL or FORTRAN is relatively easy to learn. Practice and experience are required to become proficient, but it is possible to learn to write simple programs within a matter of hours. More difficult is the efficient designing of the computational procedure itself, or *programming* proper. Some types of calculation may be perfectly sound from a mathematical point of view, but unsatisfactory in practice because of the limitations of the programming language (and indeed the limitations of the actual computer that is available). A detailed knowledge of how to code a set of arithmetic instructions is often invaluable in facilitating the choice of computing technique. Many scientists prefer, when possible, to employ specially trained staff to code their problems, but the responsibility for the overall structure of the computations is still theirs. For this reason there is considerable scope for the mathematician programmer who can design the general pattern and sequencing of computations as well as make the best use of the facilities of the appropriate language.

This mathematician programmer is often called a *program analyst*, and he is the link between the scientific user of the computer (if he is not such a program analyst himself) and the program coder. In scientific applications, the analyst may be referred to as a *scientific programmer*. The latter must have considerable knowledge of numerical analysis, as well as being well versed in various types of programming languages.

6.4 SCIENTIFIC COMPUTING

As we have seen in the last section, one of the outstanding features of an electronic computer is its capacity for the high-speed processing of prodigious quantities of arithmetic. It is natural therefore that numerical work formerly carried out on desk calculating machines is now being handled to an increasing extent by automatic apparatus. Moreover, many problems that were previously too lengthy or complex for numerical investigation can now be attempted in a practical way. We have already commented on the fact that, provided the relevant computer programs are properly tested and made to incorporate adequate checking procedures, the execution of calculations is not only several orders of magnitude faster but more accurate and also capable of much greater scope and flexibility. Nevertheless, a certain degree of caution is required if some of the difficulties arising with desk computers are not to appear in an intensified form.

The vast majority of primarily scientific problems that require special computing of some kind involve a range of mathematical calculations usually included under the title of *numerical analysis*. This covers techniques for interpolation, curve fitting, numerical differentiation and integration, the summation of series, the solution of algebraic and transcendental equations, the solution of ordinary and partial differential equations, the calculation of correlations and multiple regression coefficients, the statistical evaluation of complicated experimental results, and so on. All of these subjects involve considerable opportunities for error if the calculations are embarked upon by someone who does not have a certain minimum level of mathematical ability and insight. Just how high or low this level must be set to avoid unreasonable risk of error is a matter of fierce debate. In the past a great many scientists of modest mathematical competence have learned to carry out all kinds of difficult numerical computations for themselves without exciting much adverse comment from professional mathematicians. But with the growth of large and expensive electronic computers, and the establishment of highly organized computing laboratories, there is often great personal involvement and vested interest in numerical computation on the part of the staff concerned. This aspect of the controversy tends to obscure the fact that there are real dangers in the performance of complicated numerical calculations, whether a desk machine or a large electronic computer is used.

Suppose, for example, that the solution of some practical scientific problem seemed to depend, on the basis of certain probably erroneous assumptions, on summing the infinite series $1, \frac{1}{2}, \frac{1}{3}, \frac{1}{4}, \ldots$. In the absence of specific mathematical knowledge about the precise conditions under which such a series will converge to a definite finite sum, a perfectly reasonable approach is to continue adding successive terms until new additions appear to make little difference. The series would seem to converge rather slowly at first, and if we took 100 terms the sum would be about 5.19 with the last term added contributing only 0.01. Still, it might be noticed that adding a further 10 terms, each of which is not much less than 0.01, would increase the total by nearly 0.1. Performing these calculations, a little laboriously it is true, on a desk machine would easily lead an observant research worker to suspect that there was something wrong. On asking for mathematical advice he would quickly learn the well-known fact that the series is in truth divergent, and the sum can be made as large as we please by taking a sufficient number of terms. From this knowledge he would probably conclude that the original assumptions were in some way unsatisfactory, and these would then be scrutinized and revised. But what happens if an automatic computer is programmed to sum such a series? It is very common for textbooks to recommend stopping when the last term added is less than some arbitrarily chosen small quantity. Even with a genuinely converging series this rule may be unsatisfactory if the convergence is very slow. But if the series is actually divergent, the computer will still produce a finite number and be seriously misleading since no result exists at all in finite terms. Usually, of course, the series requiring to be summed converge fairly quickly and no difficulties arise. There would, for instance, be no trouble with $1, -\frac{1}{2}, +\frac{1}{3}, -\frac{1}{4}, \ldots$. But it may be that in crucially important cases just the kind of error described turns up.

These pitfalls are essentially mathematical. Lack of the requisite mathematical knowledge or failure to apply it may lead one to fall into the traps. More complicated examples abound. For instance, standard formulae for the numerical integration of a given function may fail to work because the function has a singularity in the range of integration. Often this difficulty can be overcome by some algebraic rearrangement or transformation before the numerical work is undertaken. Similar problems frequently arise in the solution of differential equations.

Another potent source of error derives from the necessity to represent many numbers in decimal form to an accuracy depending on the capacity of the machine in use. This leads to a round-off error of perhaps as much as half a unit in the last decimal place. While this may not appear to be serious for any individual number, especially those comprising initial values or data, it is quite easy for a serious loss of accuracy to ensue during the course of subsequent calculation. Most people are aware of the dangers of subtracting a number from another very close to it. Thus if the numbers 100·02 and 100·03 are accurate to five significant figures, the difference 0·01 may involve an error of one whole unit in the second decimal place and contain no accurate significant figures at all. Great care is needed in the planning of calculations to avoid the loss of significant figures at intermediate stages, and the accumulated effect of round-off error in a long series of computations can be catastrophic. In calculations performed on a desk machine an alert and experienced operator should be able to notice the principal sources of trouble as they occur and either make due allowance or refer the matter back to the originator of the problem. But operators of the requisite skill are often in short supply, and may in any case become fatigued and less watchful if the work is protracted. With automatic computation the intermediate steps cannot in general be examined, and appropriate safeguards must be built into the program of instructions. To construct a really foolproof program is often difficult and time-consuming, but once available such a program can be used as often as we please without fear of further errors occurring.

The necessity of having to truncate most decimal numbers may on occasion lead to results that are less obvious in practice and more spectacular in consequence. Thus, to follow an example of J. H. Wilkinson, consider the polynomial

$$(x-1)(x-2)(x-3)\dots(x-20) \equiv x^{20} - 210x^{19} + \dots.$$

The zeros of this polynomial, i.e. the solutions of the equation given by setting the polynomial equal to zero, are the integers 1, 2, 3, . . ., 20. These values must therefore result from any mathematical technique for finding zeros applied to the polynomial form on the right-hand side, at least if we can perform calculations of absolute accuracy. But suppose that the coefficient of the second term is subject to a small error, so that instead of 210 we have $210 + 2^{-23}$. If we now proceed to extract the zeros of the modified polynomial, again assuming absolute accuracy, we naturally expect to find that the new zeros are not exactly equal to the original set of integers but that the discrepancies are at least very small. It is accordingly something of a shock to learn that the original integral zeros 16 and 17 are in fact replaced by a complex conjugate pair, approximately $16{\cdot}7 \pm i2{\cdot}8$, where the imaginary parts are of appreciable magnitude. When one reflects that in most practical machine computation all coefficients might be subject to small errors, depending on how the polynomial had been arrived at, one realizes that at least in special cases the possible magnification of small errors can be of enormous extent. Moreover, we have ignored for simplicity the rounding errors introduced in practice by the numerical extraction process itself.

Other types of instability may arise from the method employed rather than from the inherent properties of the problem itself. Clearly, the previous example of a special polynomial falls into the latter class. But care is also needed in the choice of calculation, and it is often essential for sufficiently accurate results to calculate the zeros of a polynomial in a particular order. Again, the well-known Runge-Kutta method of integrating certain types of ordinary differential equations may turn out to be unreliable when the solution involves negative exponentials or even trigonometric functions. Several different kinds of instability are possible, and whether these can be avoided in a particular problem by a suitably chosen technique of evaluation can be determined only by having an adequate knowledge of numerical analysis.

While it is obviously desirable for all proposed computations, whether a desk machine or computer is used, to be subjected to scrutiny by a qualified numerical analyst, this is undoubtedly a counsel of perfection unattainable in practice. The solution seems to lie in the direction of providing all scientists who may perform their own calculations or write their own computer programs with at least enough knowledge to recognize when they need specialist advice.

It seems likely that really awkward problems do not arise quite as frequently in actual scientific work as mathematicians well acquainted with pathological examples are inclined to suggest. In many cases the accumulated rounding error can be shown on statistical grounds to achieve something like its maximum effect fairly infrequently. Again, the occurrence of grossly inaccurate solutions, such as those resulting from various types of instability, will often be easily spotted by an obvious lack of correspondence with physical or biological features of the original problem. Indeed, practical intuitions based on close acquaintance with a real-life situation can frequently be relied on to foresee the sort of mathematical difficulties that might arise. This appeal to general scientific insight should not be underestimated, and its validity is probably the reason why scientists fall into massive numerical error less often than might otherwise be expected. At the same time it is obviously highly advantageous if a scientific research worker can have the close cooperation of a professional mathematician who is personally interested in the subject matter being investigated.

For a good introduction to the elements of numerical analysis with special reference to the needs of scientists who expect to make use of automatic computers, the reader should consult the book by Hamming (1962).

6.5 COMPUTERS AND RESEARCH

It hardly needs saying that the advent of modern computers has had an enormous effect on the development of research in all scientific subjects, though only a fraction of the full potentialities has so far been realized. The foregoing sections of this chapter have dealt in a general way with the broad principles of automatic computing, as well as with the different kinds of problem that can be handled and the programming methods by means of which information and instructions can be communicated to the computer. The most obvious consequences of the computer revolution are that numerical calculations can be performed very much faster and more accurately than before, and huge quantities of data can be stored, sorted and analysed, so that whole areas of research can now be opened up that were simply inaccessible before. Moreover, the very existence of computing facilities has in turn important repercussions on the choice of research methods and even on the choice of problems worth investigating. The reason for this is that the computer not only carries out existing tasks more expeditiously, but also creates a whole new set of psychological attitudes in its potential users.

Let us first look at the narrower question of the choice of techniques in a given context. Suppose we are studying the behaviour of epidemic processes, and are looking at the usual continuous-time model for a simple stochastic epidemic with infection being transmitted in a probabilistic way from infected persons to susceptibles, but with no recovery. We can represent the process by a set of differential-difference equations for the basic probabilities. These equations are soluble by Laplace transform methods in an exact way, though the solutions for the individual probabilities are liable to be very complicated, and are difficult to handle mathematically. Numerical calculation of the probabilities, and the corresponding stochastic means and variances, from these solutions is for various technical reasons difficult to do in a satisfactory way for moderately large groups of susceptibles. The question therefore arises as to whether an alternative mathematical approach would be better *in view of the computational facilities now available.*

If we are going to compute probabilities, stochastic means, etc., anyway, we might as well attempt to do this directly from the set of differential-difference equations instead of first looking for algebraic solutions. The set of simultaneous ordinary differential equations can in fact be solved numerically by standard methods of the Runge-Kutta type. If we have an epidemic starting in a group of 50 susceptibles with one infectious case introduced from outside, there will be 51 simultaneous equations to be solved. This sounds all right in principle for a reasonably efficient computer, but great care would be needed in ensuring the validity of the numerical analysis, especially as Runge-Kutta methods are liable to be unreliable when solutions consist of sums of negative exponential terms (which they do in this case). Some alternative method of numerical integration might be better, but so far as the writer knows this particular approach has not actually been tried in practice.

A different method of computation is available based on the fact that the probability-generating function for the probabilities satisfies a certain second-order partial differential equation of parabolic type. This can be solved numerically by the fairly satisfactory Crank-Nicholson technique to give numerical values of the probability-generating function at large numbers of points in the domain of integration. Unfortunately, numerical differentiation, a notoriously unreliable process, is required to yield actual probabilities or means, etc., and the whole computation is rather time-consuming. It so happens that sufficiently accurate values of the stochastic mean are obtainable, but it must be remembered that we are considering only a simple epidemic with no recovery.

Neither this method, nor any of the others already mentioned, seems likely to be of much value if applied to more realistic models in the present state of development of numerical analysis and computer technology.

The methods discussed so far are quite feasible for very small groups of susceptibles, but then the algebra for such cases can be managed anyway. What else can we do for large groups? One possibility is the simulation approach of Section 4.5. There are various ways of doing this. We might for instance simulate the continuous time process directly. Alternatively we could use a discrete analogue of chain-binomial type, and then simulate this latter discrete process. This is in fact remarkably easy to do. The computer program required is not difficult to write, computations of an adequate order of accuracy can be performed relatively quickly on any small computer, and the method is easily extended to more complicated situations. Indeed, it has proved possible to obtain without much trouble quite illuminating information about the behaviour of epidemics in groups of individuals spread out in space over a two-dimensional area.

Thus in this kind of epidemiological problem the simulation approach not only appears markedly superior in many respects to other methods, but is also capable of being extended to deal with more complicated and realistic models than those with which we started. New vistas of research possibilities are opened up. The likelihood of being able to investigate epidemics in populations having both social structure and spatial distribution now seems considerably less remote than when one thought in terms of a purely mathematical treatment of the kind of model envisaged. Computer technology has in fact interacted with standard mathematical approaches and has pointed the way to more fruitful methods of investigation. Moreover, this is likely to provide fresh stimulus to the advancement of primarily mathematical work, so that the different modes of attack are mutually reinforcing.

Further examples of this influence of computers on research could be drawn from almost any branch of science. The implications for biology and medicine are likely to be particularly far-reaching, and indeed remarkable strides forward have already been made over the whole spectrum from molecular biology to the large-scale organization of scientific resources.

Apart from the detailed influence of computers on the fine structure of research there is the broader issue of the general climate of opinion. On the whole, anyone who really appreciates that a computer is simply a new kind of tool, providing an admittedly dramatic increase in Man's abilities to investigate and control the real world, will regard the new potentialities with optimism and enthusiasm. Such an individual will not only welcome new computer developments, but will want to know exactly how they affect his own work and how to take full advantage of them. He will also be keen that as many other people as possible should be able to understand and use the new facilities. There is often surprising opposition to the spread of new ideas and methods owing to ignorance and misunderstanding. The general public fears the development of new forces which are misrepresented as mysterious and incomprehensible, the trade unions wrongly see new threats to their workers' standards of living, and the technical experts have no wish to give up their privileged positions as high priests of a new mystique. The whole spirit of the present book is in direct opposition to such tendencies where they exist. The writer believes that it is of the greatest importance to society that as many people as possible should understand what computers can do, and should have opportunities of utilizing such equipment at first hand whenever there is a chance to do so.

These views have considerable bearing on questions like the respective roles of large and small computers. Large computers tend to be faster, more versatile, will process greater volumes of data, can handle more programming languages, can sequence whole series of computing jobs more efficiently, and so on. When the computing system is efficiently operated these gains are very considerable. At the same time there is loss of contact between man and machine. An expert mathematician who knows what a computer can do may be quite satisfied to write his own program and then leave the technical details to the machine operators. An applied scientist on the other hand, accustomed to laboratory work, will not feel happy about a remote piece of apparatus operated by people whose work he does not fully understand. But his attitude will be quite different to equipment that he can see, walk round, operate himself, and treat just like any other complex device in his own laboratory. For such reasons it is very valuable for small computers to be widely available so that they can be closely integrated with the activities of their users. As knowledge about computers spreads this aspect may become less important, but at the present time it is vital. This is especially so in medicine and public health, where the whole business of encouraging the spread of mathematical methods needs to be treated sympathetically. If clinicians and public health workers can themselves become directly acquainted with small computers the implications for their research will be assimilated at a far greater rate. Of course, the small computers will normally be run by people who are skilled in computing, but with a small installation it is possible to arrange that these people are more or less directly involved in the research of the scientists themselves.

Although the psychological effects of bringing scientific workers into immediate contact with a computer are often quite remarkable, there are a number of other specific advantages in a small installation. Thus programs requiring a lot of development are frequently handled more expeditiously on a small local machine. To begin with, complicated programs may easily have several mistakes to be rectified. Again, when a correct program is run, it may be desirable for results to be available for discussion straight away so that a modified program can be proposed and new results obtained. Where there is a long series of alternating research discussions and program modifications, more or less immediate access to a small local machine is nearly always preferable to attempting to use a large distant machine. When remote consoles linked with large central machines are freely available the situation is different, for the console is equivalent to having a machine in one's own office or laboratory.

In the meantime, it is a common experience that research requiring a lot of program development can be seriously delayed owing to long periods of waiting for programs to be processed by a big central machine. One of the main reasons for such delays is that the operators of large machines feel that the enterprise will not be economic unless the machine is run at something like full capacity, that is the 'traffic intensity' is near unity. It is well known from queueing theory (see Section 7.3) that average waiting-times are liable to be infinite if the traffic intensity exceeds unity, and may be still very high if the latter is only a little less than unity. For example, in a simple queue with random arrival of customers and negative exponential distribution of service-time, the average waiting-time is given in equation (7.6) as $\rho\mu/(1-\rho)$, where ρ is the traffic intensity and μ the average service-time. If ρ is 95 per cent, say, this means that the average waiting-time is 19 times the average service-time. Now the effective service-time of a computer installation is not just the few microseconds a computation may take to be performed on the machine, but includes all the clerical and administrative processing involved, as well as the time taken to transmit the program and data from the laboratory of origin to the computer outfit. This total service-time can easily amount to several hours unless special steps are taken to reduce it. With near-unity traffic intensities, the average waiting-time may well be measured in days rather than hours. Increasing the operational efficiency of computer service installations is clearly a matter of great importance, not to say urgency in some quarters.

For reasons such as the above, many research groups of sufficient size prefer to have their own small computer, which can be operated at a traffic intensity well below unity, and can provide very quick access and rapid processing of programs. Moreover, an independently administered machine is far more flexible and useful for practical applications, particularly for those of an operational research nature. At the same time, the availability of a really large, high-speed computer may still be essential for dealing with the more complicated problems, and for producing results of a greater accuracy than could be obtained on the small machine.

6.6 THE POCKET COMPUTER REVOLUTION

New advances in the computer world are apt to occur so fast nowadays that, before we have completely adapted to any one stage of the technological development, the next stage is upon us requiring new adaptations and new methods of thinking. Of course, looking back, we can pick out the highlights. The first big step, as already mentioned in Section 6.1, was the advent of the early thermionic valve machines around 1950. Five years later transistors were beginning to replace valves, and printed circuits were also appearing. This led to a big advance in miniaturization. Computers became smaller, more reliable and more versatile. The typical design employed a unit-assembly approach with one cabinet containing the central, basic computing unit, while several other pieces of peripheral equipment for input, output, or storage were housed in separate units.

The next big step forward occurred with the microminiaturization stimulated by military needs and space-vehicle requirements. This advance had been made possible by the development of integrated fabrication, in which complete electronic circuits could be manufactured from one small lump of raw material, instead of having to make and wire up individual transistors, semiconductors, resistors, etc. By about the mid-1960s desk-size mini-computers were beginning to appear, and in 1972 the first pocket-size computers were coming onto the market. One of these, the Hewlett–Packard HP-35, contained the equivalent of some 30,000 transistors. The pocket computers have undergone a tremendous expansion in sales, with a resultant decrease in prices. Many schoolboys can now afford the simplest calculators, and one finds such machines wherever there is arithmetic to be done quickly, simply, and cheaply.

Medium-priced pocket computers have now completely superseded the old electric desk calculating machines, and they are moreover cheaper, faster and more efficient. Special functions like square roots, powers, logarithms, exponentials, trigonometric functions, etc., are all built into the computer so there is no need to look these things up in special tables, with all the cost in time and labour that this means. In addition, statistical indices like averages and standard deviations can be obtained automatically, just by pressing the appropriate buttons immediately after the relevant series of numbers has been fed in. Opportunities for error are also very considerably reduced. Of course, compared with the large sophisticated computers, the pocket machinery is relatively slow and it has its limitations. It can, for instance, be inconvenient for handling large amounts of basic data. But for a vast range of mathematical and statistical work it is ideal. Especially is this true when one wants to avoid the administrative complications and shortcomings of a big installation, and prefers to interact directly with work in progress so as to take immediate advantage of any significant intermediate results that are noticed along the way. Finally, some of the more expensive pocket computers are programmable, just like the big computers, and this is turn vastly increases the range of feasible operations.

The earlier sections of this chapter were originally written ten years ago. With minor updating here and there, they are still essentially valid—except for the revolution in overall attitudes to scientific computing that is now taking place, consequent on the general availability of the pocket computers discussed above.

In recent years the very efficiency of modern computing methods in handling large quantities of data and in printing out innumerable tables has led to a self-negating intensity of computerization. We have already touched on this problem in Section 4.6 when dealing with medical and health statistics. Simply *because* the computer is there, people have tended to plan ever-larger surveys with bigger and bigger questionnaires, collecting astronomical amounts of data that were destined to be reissued in mountains of summary printouts. Presumably, it has been felt that the existence of the advanced technology somehow guaranteed that valuable results would inevitably be obtained. This, however, is far from being the case.

As we have seen, careful thought and efficient statistical design are essential to well-conducted experiments and surveys. With poor planning it may be impossible to achieve any worthwhile results, no matter how big the operation mounted or how powerful the computer employed. This illustrates the now all too familiar principle of 'garbage in, garbage out'. Moreover, total computing costs are soaring, although unit costs of actual calculation on the central processing equipment are decreasing because of progressively greater efficiency. But we have to include all the extra card-punching and programming time required as well, plus the administration entailed, and over and above this is the additional time spent by statisticians and scientific investigators in trying to evaluate all the summarized information available. These difficulties are further compounded by the fact that many investigations involve several years of data collection. Disillusioned statisticians find out, too late, that they are committed to trying to make sense out of assignments initiated by imprudent enthusiasts for massive data-processing activities.

This truly is a state of crisis in the use of computer facilities to which much medical and health statistical work has fallen prey. For the most part, existing commitments have to be honoured, so far as available staff and finance permit. It is of course true that resources could have been used more efficiently and that there was in principle no real need to embark so recklessly on vast data collection enterprises. Nevertheless, human nature being what it is, the temptation seems to have been irresistible. The revolution brought about by the new, cheap pocket computers does not primarily provide us with a new tool (although it does do this superlatively well), but rather presents us with a psychological opportunity to review the whole position, and consider whether there are not better ways to handle large-scale medical and health investigations.

Scientists have always, and rightly so, tried to make the most of their data, extracting the last bit of information wherever possible. Experimentation is liable to be costly and time-consuming. On the other hand, the law of diminishing returns applies. Beyond a certain point, further statistical analysis may itself be expensive and unprofitable. It must be admitted, of course, that a new analysis of old data may reveal something that had been previously missed, but this is a separate aspect.

If we are testing a long series of true null hypotheses, each being subjected to an appropriate significance test, using a 5 per cent significance level, for example, then 5 per cent of true null hypotheses will be rejected purely by chance. In these cases we shall think we have discovered something new, when we have merely observed relatively uncommon statistical events (see Sections 4.2 and 4.3 for further discussion of statistical inference). This difficulty can sometimes be avoided, if we have decided on the series of tests *in advance*, by suitably adjusting the significance level used.

Again, we may succeed in accounting for awkward data that defy an elementary explanation simply by being sufficiently ingenious in constructing an elaborate model that appears to fit well enough. Whether such a model has any real power in terms of insight into underlying mechanisms or ability to predict future events is a key issue.

In both of the situations mentioned in the two previous paragraphs the real answer to the problem of false leads lies in further experimentation or observation. Any ideas, hypotheses, models, etc., apparently supported by past material must be validated by new empirical testing. If they are mere artefacts of repeated statistical testing or ingenuity in model-building, they will in all probability be rejected by fresh data.

These dangers are of course present in any intensive and protracted analysis. They are merely exacerbated by the use of high-powered computing, unless we are specially careful to avoid the pitfalls. The situation is also liable to be compounded by the fact that excessive computer activity often leaves insufficient time for proper thought and evaluation. Statisticians and project investigators can be kept so busy marshalling the piles of summary tables coming out of the computer and making preparations to meet deadlines, such as the production of written reports, or presentations to conferences, that genuine statistical analysis on a scientific level is hardly done at all.

If this disastrous trend is to be reversed (and it does seem to have a stranglehold on many areas of investigation), far more careful thought and planning of enquiries needing statistical data will have to be insisted on in the future. One obvious point is that it is desirable to make projects as small as possible, having regard to the information required. Thus, a series of small experiments or surveys should be chosen in preference to one large exercise. On the one hand, this means one can take advantage of small computers, handling limited volumes of data but having great potentialities for detailed analysis. And, on the other hand, a series of practical investigations provides repeated empirical checks on the validity of the whole exercise.

When a large investigation is, however, unavoidable, it is preferable to use a big centralized data-processing facility only for handling the basic material. If large numbers of exploratory statistical tests need to be performed, the computer should be programmed to do all the work internally, without printing everything out—only results or tables of special interest should be printed. In this way, massive and indigestible quantities of summary tables can be avoided.

With proper advance planning this preliminary analysis of the basic data can usually be done in such a way as to facilitate scientific discussion within the investigatory group, and avoid the latter being swamped by irrelevant information. Further, it is usually possible to arrange things so that much subsequent mathematical and statistical work can be based on a smaller number of output tables, which can then be handled independently on a pocket computer.

Another aspect of dealing with investigations that may turn out to be large is to plan for a connected sequence of stages, with careful evaluation at each stage so that the whole enterprise can be easily terminated at an early point if required. A typical sequence would involve a preliminary study to make a 'situational analysis' of existing knowledge, relying mainly on library research. This might take only two or three months, and need only one or two people.

If a larger experimental or field study seems to be indicated, it is often advisable to carry out some simple preliminary theoretical analyses, perhaps using a small computer for simulation purposes. These analyses should be directed towards determining the key features on which a full-blown investigation should concentrate. If such an investigation seems clearly justified, then the relevant planning must be undertaken, and subjected to critical review. Finally, before embarking on the major study, which might take a number of years to complete, and involve much time, effort, and expense, a small-scale pilot study should be mounted first in order to test the feasibility and accuracy of the techniques proposed.

Most of the preliminary work may well be within the capabilities of a pocket computer. Also, because of the careful planning and gradual build-up, the final investigation is likely to be much more efficiently tailored to deal with essentials. It may be that all the subsequent work can be done on a pocket computer, though this will not be possible if the amount of data to be handled is very large. But, in any case, every effort should be made to use a large computer installation as little as possible—concentrating on the minimum necessary data-handling aspects. After this, small computers can be used to carry out the main scientific and statistical calculations on a decentralized basis. A further advance may well occur before long: this would be for the output from a large computer, regarded as essential for the basic handling of a large data base, to be recorded in such a way that it could be fed directly into a pocket computer for further analysis. Again, new methods of microminiaturization may enable even large data bases to be handled in a physically very compact form. This would mean that a complete computing centre would be reduced to desk or pocket dimensions.

The above schemes may be somewhat idealized, but if efforts were made to design enquiries along the general lines suggested, it is my conviction that enormous savings could be made in time, staff, and finance, and that considerable increases in scientific and practical utility of results would be obtained.

CHAPTER 7

Operational Research

7.1 GENERAL SCOPE AND DEFINITIONS

Although the scientific approach in general, and mathematical methods in particular, may be applied to almost any field of endeavour concerned with the classification and understanding of facts, it is convenient to make certain distinctions. There is, for example, the traditional dichotomy of pure and applied science. No hard and fast line can be drawn, but it is broadly true to say that pure science is mainly concerned with the understanding of fundamental phenomena, while applied science is occupied with the practical utilization of this knowledge to control and manipulate our environment. In both cases the scientific content, i.e. the physics, chemistry, biology, etc., is of basic importance and is a matter of explicit study.

However, in making the benefits of applied science available to the community at large, all kinds of additional problems of economy, efficiency, and organization are involved. Industrial plant investment, distribution, marketing, advertising, information services, public and private planning, etc., are just some of the aspects to be taken into consideration. Many of these activities are more concerned with the organization of resources, both human and material, than with the basic scientific or technological nature of the resources themselves. In short, it is the problems facing executive and administrative authorities that are here being emphasized. This whole field is one which may very properly be made the subject of scientific inquiry and investigation. It is convenient to refer to it in a general way under the heading of *operational research* or *operations research* (the former usage being British and the latter American). The definition of operational research as the scientific study of executive or administrative problems is certainly very valuable in underlining important differences from both pure and applied science. Parts of this whole area have long received special emphasis under such designations as time and motion, work study, organization and management, scientific management, human engineering, etc., and more recently there have been newer disciplines such as ergonomics and cybernetics. It is not part of our purpose to try to define and catalogue all these different subjects, some being quite distinct and some overlapping, some being characterized by the field of application and others by the techniques employed. The main task of the present chapter is to draw attention to a large and rapidly developing collection of scientifically oriented activities that are concerned more with forms of organization than with the actual matter being organized. Naturally, the two aspects of form and content interact, and an adequate knowledge of both is required for successful practical applications. Operational research methods are of growing value in the practical utilization of theoretical biological knowledge, and are already of established importance in optimizing the organization and distribution of medical care.

One frequently used phrase which we have not introduced so far is *systems engineering*. The actual usage in the literature is extremely variable: some writers regard it as entirely synonymous with operational research, while others draw clear distinctions. It is possible to say, for example, that operational research workers are more concerned with operational procedures, particularly those that already exist, whereas the systems engineer thinks more of a whole man and machine complex, especially with regard to the design and development of new systems. Again, precise usages vary with different writers. There is often some advantage in being able to distinguish the kind of study that is concerned with the organization and functioning of a whole system of processes from investigations into the detailed operation of any component part. Thus we could label the examination of appointment systems in hospital outpatient departments as operational research, reserving the term 'systems engineering', or perhaps *systems analysis*, to consideration of the totality of medical services in a large area or over the country as a whole. There is perhaps no special logical advantage in maintaining such a distinction, and we could regard all scientific approaches to the functioning of a system as coming within the purview of operational research. However, there are some practical advantages in discussing the behaviour of very large, complex systems of interacting components separately from more immediate, down-to-earth, applications to limited areas. We shall, accordingly, reserve discussion of systems analysis, and the associated topic of system dynamics, until the following chapter.

Administrative and executive problems tend to have a good deal in common, irrespective of their actual subject matter. Thus subscribers waiting to make a telephone call, customers waiting in several lines for service in a bank or post office, aircraft stacked over an airfield waiting to land, cars waiting in a traffic jam at a bottleneck, and patients waiting on a waiting list to enter a hospital inpatient department or in a waiting-room for an outpatient consultation for which they have been given an appointment, all have in common the fact that some kind of service has a certain pattern of demand and a certain pattern of supply, and the interaction of these two patterns results in a fluctuating queue or waiting-line. The mathematical *theory of queues* and allied processes is accordingly of cardinal importance to any operational research investigation involving some sort of supply and demand. Again, many industrial processes have in common certain problems of optimization in respect of the output of a range of products. For instance, an oil refinery may entail a cracking process which permits various possible outputs of petrol, paraffin, fuel oil, etc. Suppose we know the relationships between the outputs of the different products, together with their market prices and the costs of raw materials. Then the basic question arises of determining the optimum mixture of allowable outputs which will maximize the firm's total profit. Of similar theoretical form is the transportation problem of determining the minimum cost pattern of distribution between a given number of factories and warehouses. These optimization problems can usually be handled by some *mathematical programming technique* such as *linear programming*.

There is in fact a whole range of mathematical methods that are of special value in operational research, and many of them have moreover been largely developed because of their applicability to that field. A more extended discussion of some of these theoretical techniques is therefore given in Section 7.3.

For detailed survey of the whole subject of operational research and systems engineering, the reader is referred to the excellent book by Flagle, Huggins, and Roy (1960), while more specific discussion of various mathematical techniques and their applications may be found in Churchman, Ackoff, and Arnoff (1957), or Sasieni, Yaspan, and Friedman (1959).

7.2 ORIGINS AND DEVELOPMENT

The origins of operational research can certainly be traced back as far as the work of Frederick W. Taylor, on what he called 'scientific management', in the United States during the early years of this century before World War I. There were many advances between the two world wars involving accounting techniques, time and motion studies, statistical quality control, etc., which were all relevant to improving the quantitative aspects of managerial decision-making. At the beginning of World War II, however, there was a recognizable jump forward in the conscious control and handling of what rapidly became generally known as operational research. It was evident that supreme importance attached to reaching optimum decisions as rapidly and as efficiently as possible on a variety of military problems involving radar systems, anti-aircraft gunnery, choice of convoy size, the planning of bombing raids, antisubmarine warfare, etc. In each of these spheres there was of course a basic emphasis on trying to utilize the best scientific and technological knowledge available. But given any existing piece of apparatus, equipment or machinery, there clearly remained the vital problem of how best to handle and organize it. A group known as 'Blackett's circus' was assembled under the leadership of Professor P. M. S. Blackett, and consisted of a broadly based team that included several mathematicians, physicists and biologists, as well as an astrophysicist, a surveyor, and an army officer. The multidisciplinary character of this team was essential to its success, ensuring that attention was never exclusively concentrated on any single scientific or technical aspect of a problem, and enabling conflicting claims and requirements to be properly considered and appreciated. Blackett's group dealt with many of the problems mentioned above, and it has often been regarded as the prototype of the modern operational research team.

As an illustration of the application of the operational approach to a specific wartime problem let us look at an example with which the writer was personally concerned. It had long been customary for the probable position of a radio transmitter to be estimated from directional bearings taken on suitable signals from convenient points. If the bearings were without error they would all pass through the actual location of the transmitter, and no more than two would be required. Unfortunately, such bearings, especially when made on high-frequency signals, were subject to considerable uncertainty. The position of the transmitter might therefore be difficult to determine at all accurately, even when large numbers of directional observations were available. Up to a point, of course, increases in technical efficiency in the construction and operation of direction-finding apparatus helped to overcome some of the difficulties. But influences like large unpredictable ionospheric deformations were beyond technical control, and could at best be allowed for by appropriate statistical procedures in handling the data.

The conventional method of making direction-finding 'fixes' was to set up the bearings on a gnomonic map projection, on which they would appear as straight lines and could therefore easily be represented by lengths of string suitably secured at each end. An intuitive appraisal of the 'bird's nest' of lines on the chart would then be made with a view to estimating the most probable position of the transmitter along with some indication of how accurate this estimate was likely to be. Paradoxically, perhaps, it is an unavoidable feature of such visual methods that, although they may be surprisingly accurate with small numbers of high-grade bearings, they become increasingly difficult to operate efficiently as the number of bearings becomes larger and a more confusing picture is present, although the amount of inherent information is in fact correspondingly greater.

By constructing an appropriate mathematical model of the process it proved possible to develop a computational method by means of which the most probable position of the transmitter could be calculated together with a rectangle within which there was some assigned probability, say 90 per cent, of the transmitter's lying. Specially designed maps and mathematical tables enabled the necessary arithmetic to be performed fairly quickly, even for large numbers of bearings, using only an ordinary desk calculating machine. Nowadays, such calculations can be handled far more rapidly and accurately by using an electronic computer.

To begin with, computed 'fixes' on transmitters of known location were compared with visual determinations, to assess the relative accuracies of the two methods. As the operational use of the mechanized technique progressed, it was found that there were many situations in which it could be used to advantage in its own right. On the other hand, there were also important occasions when the available data were ambiguous or confused in some way, or when snap decisions were urgently required. Individual judgement on the part of the direction-finding plotters then had to be relied on, but this judgement had now been improved and refined by contact with the computational technique. There was thus a fruitful interaction between human intuition and mathematical technique: each could contribute something to the development of the other. A somewhat analogous situation arises in connection with computer-assisted medical diagnosis, which was discussed in Section 4.7.

Parallel developments of operational research also took place in the United States during the war, involving such matters as antisubmarine warfare and the effectiveness of the Air Force. Moreover, operational research organizations have continued to flourish since that time in all departments of the American armed forces.

Since the war there has been an enormous increase in the application of operational research methods to a great variety of industrial and social organizations, such as railways, hospitals, supermarkets, newspapers, coal and electricity supplies, building design and construction, etc. The extent of the applications and the enthusiasm with which they have been made have been considerably greater in the United States than in Great Britain, although notable achievements have occurred in both countries.

It is obvious that there must be great scope for the use of operational research techniques in the application of biological and medical knowledge to human problems. In subjects like forestry, agriculture, and fishery, there is obviously a large element of executive and administrative activity. The organizing of scientific understanding for the purpose of controlling silviculture, farm management, or hatchery policy in an optimum manner presents difficulties which are not covered by the mathematical methods normally used in pure and applied science. Similarly, the design and organization of hospitals and the efficient distribution of medical care generally, involve problems which are no part of medicine as such. Considerable strides have been made in the adaptation of operational research techniques to the needs of the medical services and health planning, and Sections 7.5 to 7.7 are devoted to a detailed discussion of some of the work that has been done in this area.

7.3 SOME OPERATIONAL RESEARCH TECHNIQUES

Although the essential character of operational research lies in applying the scientific method to organizational problems, attention is often focused on the collection of mathematical techniques which have been found to be of special value and importance throughout most of the field. For example, the widespread occurrence of demand for and supply of certain goods or services involves problems of queueing, marshalling, sequencing, scheduling, inventory control, etc. To begin with, a mathematical model of the appropriate activity is required, together with an investigation of its properties. Then, a method must be found for calculating the optimal pattern of operation, on suitable criteria, of the actual process in the real world. Queueing theory and linear programming have already received brief mention in Section 7.1. It is proposed to give below a little more detail about a selection of topics that have attracted special interest, and which seem to be of central importance to operational research.

Queueing Theory

The salient features of any queueing situation are a continual, though often variable, demand for certain goods or services, and a service organization for satisfying this demand. The simplest example is that of customers arriving at a shop counter to purchase certain items. If the counter is free when a customer arrives, he is served immediately. But if the server is already occupied with an earlier customer, the new arrival has to wait, and characteristically has to join a queue to await his turn. This simple situation is capable of a huge number of modifications, extensions, and developments, and there is now an enormous literature on queueing theory. For lists of references and detailed mathematical discussion the reader should consult Cox and Smith (1961), Saaty (1961) or Chapter 11 of Bailey (1964). In fact, the simple set-up readily identified as a straightforward queueing situation is really just a special case of more general input-output processes. On the organizational level, the whole system of supply and demand in the context of medical care is such a process. On the biological level, the totality of metabolic activity may be viewed in this way. However, let us now return to the elementary queueing model in order to discuss its characteristic properties.

First, we envisage the *input process* as involving no more than a stream of customers demanding service. Frequently, it is convenient and appropriate to assume that the customers arrive at random, though at some fixed average rate. This means that the time distribution between successive arrivals will be negative exponential in form, and that the number of arrivals in any fixed period of time will have a Poisson distribution. Alternatively, we might impose an appointment system which entailed the customers arriving at fixed points of time. There is obviously a whole range of intermediate possibilities.

Secondly, we must specify the *queue discipline*. The simplest assumption is to use the 'first come, first served' rule. The customers thus wait in line in order of arrival, the individual at the head of the queue being the only one to receive attention at any given time. More complicated disciplines can be handled by specifying some kind of priority system, or the arrival of urgent demands, such as emergency cases in hospital, requiring instant service.

Thirdly, there is the *service mechanism* which deals with the output end of the process. Basically, we need to know the number of servers and the statistical distribution of service-time. The single-server case is naturally the easiest to deal with.

It is obvious that if customers arrived at fixed intervals and if the service-time were constant, then when the latter was less than or equal to the inter-arrival time no queue would build up, while if it were greater the queue would increase indefinitely at a fixed rate. No problems of understanding arise in such a simple deterministic case, but as soon as various probability elements are introduced a number of special features appear. Therefore, to fix ideas, let us assume that we have a single-server queueing process, with customers arriving at random with an average inter-arrival time λ, and with a general service-time distribution having mean μ and variance σ^2. If $\mu > \lambda$, common sense suggests that as the demand is outstripping the supply the length of queue will continually increase, and that a highly unsatisfactory state of affairs is in existence. Conversely, if $\mu < \lambda$, the average demand is less than the average possible supply so the situation ought to be reasonably satisfactory, except that a queue is bound to build up occasionally owing to statistical fluctuations. The precise relationship of queue length to λ and μ clearly needs examination. The special case $\lambda = \mu$, when average supply and demand exactly balance, is specially interesting and has properties contrary to naïve expectation.

It is possible to show, by fairly elementary reasoning, that the average queue length $\bar{q}$ is given by

$$\bar{q} = \frac{\rho(2-\rho) + \sigma^2/\lambda^2}{2(1-\rho)} \tag{7.1}$$

where $\rho = \mu/\lambda$ is the so-called *traffic intensity*. If the service-time is in fact constant, we have $\sigma^2 = 0$, and (7.1) takes the value

$$\bar{q} = \frac{\rho(2-\rho)}{2(1-\rho)} \quad \text{(constant service-time).} \tag{7.2}$$

Again, if the service-time is negative exponentially distributed, the variance σ^2 will be $\mu^2 = \rho^2\lambda^2$. This time (7.1) becomes

$$\bar{q} = \frac{\rho}{1-\rho} \quad \text{(exponential service-time).} \tag{7.3}$$

We can also calculate the average waiting-time $\bar{w}$ that a customer has to undergo before he starts to receive attention from the server. The corresponding general expression for $\bar{w}$ can be shown to be

$$\bar{w} = \frac{\rho^2 + \sigma^2/\lambda^2}{2(1-\rho)/\lambda} \tag{7.4}$$

with special cases

$$\bar{w} = \frac{\rho\mu}{2(1-\rho)} \text{ (constant service-time)} \tag{7.5}$$

and

$$\bar{w} = \frac{\rho\mu}{1-\rho} \text{ (exponential service-time).} \tag{7.6}$$

Comparison of (7.5) and (7.6) shows how replacing a constant service-time by an exponentially distributed one exactly doubles the average waiting-time.

It is clear of course that all the formulae (7.1) to (7 6) are valid only for $\rho < 1$. As ρ approaches unity, so the average queue length and average waiting-time tend to infinity. Thus a highly undesirable state of affairs exists even when supply and demand exactly balance. Although there is now a very extensive body of literature on the mathematical theory of queueing processes, one of the simplest and most valuable practical applications is the more or less commonsense requirement that the traffic intensity should be kept below unity. Of course, the value of the theory is that it allows one to calculate precisely how much waiting would be entailed by any given traffic intensity. The latter should be as near unity as possible to make the most efficient use of the server's time, but as low as possible to achieve the minimum amount of waiting. Hence it may be feasible to plan for an optimum balance of these conflicting characteristics if we can compute the exact consequences of different methods of practical operation.

Similar considerations apply in principle to the more complex systems such as those with several servers, queue disciplines involving special priority rules, and input processes of an arbitrary or complicated nature. Some specific applications of queueing theory in the context of balancing the demand for and supply of medical care are discussed in Sections 7.5 to 7.7.

Mathematical programming

In the majority of organizational problems susceptible to some kind of operational research investigation, economic aspects are likely to be fairly prominent. These may be more or less implicit, as in the queueing type of process just discussed where economies are sought both in customers' waiting-time and in the use of servers' time. If it is possible to attach monetary values to the loss of customers' and servers' time, an overall analysis can be made to show the net loss to the community. More generally, however, it is usually a question of making an intuitive judgement of the best balance between conflicting requirements that are difficult to quantify on the same scale of measurement.

On the other hand, many problems have the economic aspect explicitly built into the basic formulation. Let us consider a typical manufacturing problem that has its counterparts in several other fields where the main problem is one of optimizing the allocation of resources. We suppose that the primary object of the firm in question is to maximize its profits from the sale of a range of different items, subject to various constraints. Let there be n different items, and let the output of the ith item be x_i units over some specified interval of time. Then if the profit per unit of the ith item is c_i, the total profit z is given by

$$z = \sum_{i=1}^{n} c_i x_i. \tag{7.7}$$

If any c_i is negative, the corresponding product is being sold at a loss.

Broadly speaking, we wish to find those values of the x_i which maximize z, given the c_i, and subject to certain conditions imposed by the limitations of the manufacturing process. To begin with, of course, the x_i cannot be negative. Thus one set of restraints is

$$x_i \geq 0 \quad (1 \leq i \leq n). \tag{7.8}$$

Next, suppose that the total number of man-hours available is fixed at b, and that the number of man-hours required to produce one unit of the ith item is a_i. The total resources required to produce the output $x_1, x_2, \ldots, x_n$ is accordingly

$$\sum_{i=1}^{n} a_i x_i,$$

and this cannot exceed b, i.e.

$$\sum_{i=1}^{n} a_i x_i \leq b. \tag{7.9}$$

Similar considerations may also apply to the number of machine-hours available on a particular machine, the supplies of various types of materials, the resources available for distribution and marketing, etc. If there are m restrictions in all we can represent these by a generalization of (7.9), namely

$$\sum_{i=1}^{n} a_{ij} x_i \leq b_j \quad (1 \leq j \leq m). \tag{7.10}$$

The mathematical problem, therefore, is to maximize z in (7.7) subject to the restraints imposed by (7.8) and (7.10). In this somewhat oversimplified version of a commonly occurring practical problem, the function z and the restrictive conditions are all linear functions of the x_i. The model in this case is referred to as a *linear programming model*. A variety of standard methods exist for solving such problems, but a considerable amount of arithmetic may be required in practice if the numbers of variables or constraints are at all large. Fortunately, computer programs are available for the rapid derivation of solutions.

We shall not discuss here the detailed theory of handling linear programming models, though the following geometrical interpretation may be helpful in visualizing some of the salient features of the problem. (For a comprehensive treatment of this whole subject the reader should consult one of the many textbooks now available, e.g. Garvin, 1960.) Suppose to begin with there are only two different items involved with outputs x_1 and x_2, i.e. $n=2$. All possible kinds of output can be represented in a plane by the coordinates (x_1, x_2). Since we must have $x_1 \geq 0$ and $x_2 \geq 0$, there is an immediate restriction to the positive quadrant of the plane. Next, consider the jth linear constraint $a_{1j}x_1 + a_{2j}x_2 \leq b_j$. All points for which this inequality is satisfied lie in the half-plane on the same side of the line $a_{1j}x_1 + a_{2j}x_2 = b_j$ as the origin, provided $b_j > 0$. It is easy to see that in general the solutions permitted by the restrictions all lie within a polygon bounded by the $m + 2$ lines

$$\left.\begin{aligned} &x_1 = 0, \qquad x_2 = 0, \\ &a_{1j}x_1 + a_{2j}x_2 = b_j \quad (1 \leq j \leq m). \end{aligned}\right\} \tag{7.11}$$

Finally, the function z is constant on any line $c_1x_1 + c_2x_2 = d$, taking the value $z = d$. A little reflection shows that the greatest value of z permitted, having regard to the imposed constraints, is achieved by a line which is as far from the origin as possible and just touches the polygon specified by (7.11). The determination of the precise position of this line (if a solution exists) with the associated value of d giving the maximum value of z is now largely a question of using an appropriate computational technique.

It is not difficult to generalize the above ideas to the case of n different items ($n > 2$), and to incorporate inequalities of the form $\Sigma a_{ij}x_i \geq b_j$ as well. The main task of the operational research worker is of course to recognize that a particular problem in the real world can be formulated in linear programming terms, at least approximately, so that the standard computational devices can be applied.

We have deliberately chosen an illustration above in which the function z and the associated constraints are all linear. However, in economic contexts a law of diminishing returns is frequently observed, so that increased production of any item results after a certain point in a decreased profit per unit. Again, it may happen that profits obtainable from different items are correlated in some way, and are not independent as we have implicitly assumed. If such effects are to be taken into account, a more general function must be chosen for z. When z can be expressed as a quadratic function of the x_i, then subject to certain conditions and still keeping the constraints linear we say that we have a *quadratic programming model:*

$$\left.\begin{aligned} z &= \mathrm{f}(x_1,\ldots,x_n) & & \\ x_i &\geq 0 & &(1 \leq i \leq n) \\ g_j(x_1,\ldots,x_n) &= 0 & &(1 \leq j \leq m). \end{aligned}\right\} \tag{7.12}$$

It is obvious that matters of very considerable complexity can arise both in theoretical treatment and in computational handling. Wherever possible one prefers to work with a linear programming model providing this is not adjudged to be too unrealistic.

A further complication of practical importance arises when we wish to determine the optimal policy over a period of time. Conditions may be known to vary from month to month. Although one reasonable procedure would be to adopt a linear programming approach for each month separately, it is not certain that this yields the best possible results overall: the mode of production in one month may place us in a better or worse position with regard to operations in the next month. This type of problem may be handled by using the notions of *dynamic programming*, due in large part to the work of Richard Bellman.

It can be seen from the foregoing that there are a number of powerful mathematical techniques available for dealing with a wide range of problems in which optimal policies are sought. Decisions rest in the final analysis with the administrators or executives concerned. But the important point is that a large part of the area formerly dealt with purely on the basis of individual experience and intuitive judgement can now be handled in a far more precise way, leaving human powers of insight and decision correspondingly freer to concentrate on those matters where they are most needed.

Simulation and operational gaming

The general notion of *simulation* has already been discussed in Section 4.5 in connection with the investigation of complicated stochastic processes. In so far as operational research studies must, of their very nature, be extremely close to real-life situations if they are to be successful, it follows that suitable models of the relevant human activities are almost certain to involve some highly intractable processes. On the one hand we want specific information about the general character of these processes, and on the other hand we want to be able to calculate the probable consequences of changing the values of certain fundamental parameters which may be under control. In both cases simulation may be the only feasible way of making progress. It thus comes about that simulation is one of the most important basic mathematical techniques of operational research, and it is worth saying a little more about the role it has to play.

Suppose, for example, that we want to study the operation of an international airport with special emphasis on the rates of flow of people, together with their luggage, through the whole system. We have in effect a very complicated multistage queueing process with the possibility of bottlenecks occurring at various points because of the formation of subqueues. Thus passengers arrive at the airport, mainly in accordance with predetermined flight times but with a good deal of statistical variation in punctuality. They have to check in at the appropriate airline desk, have their luggage processed, pass through customs, and wait for their flight to be announced. The plane may arrive on time, or may be late because of delays during flight or because of delays in obtaining permission to land owing to local traffic congestion. Incoming passengers go through the reverse process, but have in addition to check with the immigration authorities and have their baggage checked by customs. An adequate stochastic model of such a situation would not be too difficult to formulate, and would depend partly on a detailed knowledge of the various states and stages to be incorporated, in short the general patterning of the process, and partly on a knowledge of the frequency distributions associated with arrival and service at different points. A careful survey of the whole system could establish these matters with a sufficient degree of accuracy.

Formulating such a model is one thing, but to try to carry out a theoretical analysis would be a very different story, and would in general be almost certainly impossible. But to set up computerized simulations to study the behaviour of the whole system over a relatively long period of time would be perfectly feasible. Not only should we obtain a very useful quantitative description of the system actually in operation, but we would also be able to forecast the probable consequences of any suggested rearrangements and improvements. Very often actual changes in procedure and operation are difficult to introduce, may have to run for long periods of time to determine whether they are significantly better or worse than the previous program, and may lead to considerable loss of profit and discontent amongst staff and customers if unsuccessful. The simulated study of a suggested new mode of working enables many important issues to be settled theoretically and quite painlessly. Only if extensive investigation of this kind suggests that appreciable improvements are likely to be forthcoming should we embark on an actual operational trial. Some unforeseen difficulties may still arise, but at least they will be far fewer and less disruptive than if no simulation had been undertaken. Moreover, it would be quite out of the question to test directly more than a very small number of new arrangements as several lifetimes would be required. Using a high-speed electronic computer we may be able to obtain a simulated version of many years' real-life operation in a matter of hours, or even minutes. It thus becomes quite practicable to examine in this way a wide range of alternative reorganizations before submitting just one or two to the crucial test of empirical verification.

On a rather more abstract level than the foregoing example is the use of simulation methods in *operational gaming*, where we are more directly concerned with the way in which human decisions are repeatedly made in the face of a continuous presentation of complex problems. The airport set-up just discussed is very similar to industrial situations where materials of various kinds have to be dealt with in a purely mechanical way as rapidly and economically as possible. Given a particular pattern of organization together with its associated parameters, we can compute the numerical consequences and so obtain discrimination between alternative organizations. Once a fairly optimal system has been achieved the main problem is then to keep it operational. In contrast to this type of application we now want to concentrate on the level at which human decision making is paramount. One of the most highly investigated areas is the choice of optimal tactics and strategies in various kinds of combat situation, as developed, for example, by the Tactical War Gaming Group of the U.S. Operations Research Office.

The conduct of any sort of battle is obviously subject to chance influences to a very high degree, and the choice of an optimal strategy is often one of the utmost difficulty. Some aspects of the airport problem are present here in an intensified form. Thus in most cases it is possible to carry out only one of the many alternative strategies that can be envisaged. A practical test of alternatives is more than difficult and time-consuming: it is impossible. It has, therefore, long been the practice of battle commanders to carry out crude 'thought-simulations' in their heads, or to develop imaginary actions in terms of map exercises, with a view to choosing what appears to be the best policy. Because of the large amount of chance variation involved even the best strategy may fail, so that the problem of drawing statistically correct inferences from a knowledge of the practical results obtained by other commanders in different circumstances is enormously difficult. Nevertheless, the salient point is that successful common sense does in fact rely on a process which is essentially an elementary form of simulation. Hence it is likely to be worthwhile attempting to put the whole process on a much more quantitative basis.

One method that has been tried is to set up an artificial battle situation on paper and have opposing teams of players who themselves make decisions stage by stage on the basis of information presented to them. Thus at any particular stage it is people who decide whether to attack, retreat, open fire, call for tank support, etc., although the consequences of these decisions are determined artificially in a probabilistic way. It is naturally highly convenient and desirable to use a computer to calculate the results of decisions in an appropriate manner, to handle the recorded progress of the battle, and to produce any analyses and summaries that may be required. There are many theoretical difficulties and objections to an artificial game of this type. First, the inevitable complexity of any battle usually means that weeks may be required to finish a single game. Large numbers of repetitions are therefore prohibitively expensive and time-consuming. Adequate discrimination between alternative strategies is thus hard to obtain, although the participants may in fact learn a great deal of practical value from being forced to think more quantitatively and to reflect in a relatively unstressed situation on the outcomes of their decisions. Secondly, the fact that learning and modification of human judgement takes place during the progress of a game means that the general conditions of play cannot be held constant. This increases the difficulty of deciding between alternative strategies, or alternative weapon systems, other things being equal.

A more promising procedure is to have a mathematical model of the whole battle process, including the decision-making aspect. This is not to devalue the importance of the latter, but on the contrary to allow it to be incorporated as a major integral part of the model in such a way that it can be more satisfactorily studied. Broadly speaking, all decision processes that might occur in the battle have to be coded in such a way that they can be repeated, with appropriate statistical variation, as often as required. We can then computerize the entire action of the battle, and achieve a reduction in the total time required of several orders of magnitude. It is now perfectly feasible to carry out Monte Carlo type simulations to compare the effectiveness of different strategies and weapons systems in a statistically acceptable way. This kind of *tactical war game* has been developed by the U.S. Operations Research Office under the code name *Carmonette*. For a detailed discussion of the techniques employed the reader should consult Chapter 25 of Flagle, Huggins, and Roy (1960).

The possible uses of gaming have not yet been fully explored in relation to the application of operational research methods to medicine and public health. One field in which the decision-making process is of paramount importance is public health administration. Action is taken against a general background of clinical, biological, economic and sociological knowledge, but forecasting the precise consequences of any particular action is often difficult if not impossible. Operational gaming was applied as early as 1964 by the Systems Development Corporation and the California Department of Public Health to problems of training and research in the public health field (see *SDC Magazine*, January 1964). The basis of this work was the construction of a simulated city called 'Epiville', which was essentially an elaborate model of a hypothetical city involving as explicit features all the common activities and functions of public health personnel. Real human decision makers took part in the exercise and were confronted with a variety of problems thrown up by simulated epidemic outbreaks. Their actual performances in terms of decision making could subsequently be evaluated in joint discussions with special reference to such questions as whether they received adequate information about outbreaks, whether they made the correct decisions, whether they succeeded in bringing the epidemic under control, and so on.

There seems to be little doubt about the educational value of such gaming. Most participants reported that they experienced the anxieties and tensions associated with real-life situations, and felt a corresponding sense of responsibility. Under these conditions it is possible for public health personnel to have a wide and useful experience of dramatic situations that arise only rarely in practice.

Apart from the value in training administrators, simulated communities could be used to test out different types of organization. This could with advantage be done over the whole field of medical and health services. Indeed, any operational research attempt to study the total system of supplying medical care is likely to find simulation work essential, as purely mathematical investigations are almost certain to be hopelessly intractable.

The theory of games

Another topic that features prominently in many operational research studies is the so-called *theory of games*. There are certain important distinctions between what is usually meant by this term and the *operational gaming* of the foregoing paragraphs. In the latter, gaming is used to refer essentially to the idea of simulating some highly complex and variable decision-making process by means of a game played according to appropriately codified rules, possibly involving actual human players but preferably having imaginary players whose decisions are determined by suitable calculations. The theory of games, on the other hand, is primarily concerned with methods and principles of choosing the best play in some well-defined competitive situation. Practical instances of competitive behaviour abound in business and industry. They are for the most part too complicated for direct analysis, though some useful work has been done in the study of competitive bidding for contracts between rival companies. Theoretical studies are usually confined to highly simplified and stylized situations, but for this very reason help us to discern some of the more important aspects of real-life problems. In order to gain some idea of what is involved let us consider the following elementary illustration.

Suppose that two persons, A and B, play a certain game, the precise details of which need not be specified. We assume that the players strive to win each play within the limitations set by the rules of the game. The goal or end-state of each play is a win for A, a win for B, or a draw. At the conclusion of each play one player pays the other a certain sum of money, the actual magnitude of which depends on the way the play ended. A draw involves zero payments, but the amount A receives if he wins may depend on the nature of his victory. Now, as already mentioned, what we are really concerned with in game theory is not the details of individual steps or moves in any particular play, but the general plan of action, or *strategy*, to be adopted. Let us suppose that A knows of three alternative strategies, A_1, A_2 and A_3, one of which he might choose to adopt, while B is in possession of two possible plans of action, B_1 and B_2. We now assume that it is known what payments will ultimately be made by one player to the other if any given pair of strategies are adopted, that, for example, B wins 20 points from A if A follows A_1 and B follows B_1. The consequences of the six combinations of strategies can be exhibited in the form of a matrix which specifies the amounts paid by B to A for each pair of strategies. This *pay-off matrix* for the present illustration is shown in the following table.

		Player A		
		A_1	A_2	A_3
Player B	B_1	− 20	− 10	10
	B_2	20	30	20

It is immediately obvious that, so far as B is concerned, strategy B_2 is not a good one, as he always loses and has to make some payment to A. If he follows B_1 he wins when A adopts A_1 or A_2, but loses 10 points when A has been wise enough to choose A_3. The best strategy for B is always to choose B_1. Now A, on the other hand, does best if he follows A_2 and B follows B_2. But B is unlikely to do this, for the reasons already given. If B chooses B_1 then A's maximum return occurs when he operates plan A_3. The best strategies for A and B are therefore A_3 and B_1, respectively, resulting in B paying A ten points. This quantity is called the value of the game. When, as here, a player's best choice is to keep to the same plan at each play of the game, he is said to use a *pure strategy*.

The optimal strategies can in fact be found by means of a useful operating rule—the minimax principle, according to which it is reasonable to choose a strategy which minimizes one's greatest possible loss. Thus for player A, the greatest losses he could incur are $+20$, $+10$ and -10 for strategies A_1, A_2 and A_3, respectively. The least loss here is -10. Hence A_3 is a minimax policy. Similar reasoning applied to B shows that his greatest losses are $+10$ and $+30$ for B_1, and B_2, respectively. B's minimax strategy is therefore B_1. We thus have the same result as before. In general, if we have player A's strategies corresponding to columns of the pay-off matrix, we look for an entry that is the highest in its row and the lowest in its column. If such an entry exists, there is said to be a saddle-point, and the corresponding strategies are the best for the two players concerned.

It often happens, however, that there is no saddle-point. In this case there is no single pair of strategies, each of which is the best for the player concerned under all circumstances. The best method of procedure is for the players to adopt a *mixed strategy*, which means that each player makes a random choice between certain of the plans available to him, the probability of choice depending on the actual set of numbers appearing in the pay-off matrix. The complete solution of such problems can be fairly complicated. Several methods of procedure are available, one of which is to convert the whole set-up into a linear programming problem and use the relevant treatment of the transformed situation.

Now, although the above type of discussion is a perfectly valid approach to games in which we are primarily concerned with the choice between alternative broad strategies without consideration of the minutiae of play, it is quite possible to redefine the competitive situation so as to include these details. Thus in poker, bridge, or chess, we are interested in the best card to play or piece to move at any stage of a single game. Under certain general conditions, it can be shown that for games specified in this *extensive form* each player has an optimal pure strategy. Unfortunately the number of alternatives is usually too large for this approach to be of practical value, though the theoretical implications may be worth investigation.

We have only touched on the fringe of the subject here. Many interesting extensions are possible to many-person games, in which there may or may not be coalitions; to games involving pay-offs which are continuous functions instead of discrete as above; to games in which one person's loss is not automatically another's gain, as in economic competition between companies; and so on. The classic work on the subject is that of von Neumann and Morgenstern, first published in 1944, but a considerable literature has developed since then.

It is uncertain how much has actually been achieved by way of practical applications of game theory. Military applications have been referred to, but for obvious reasons have not been published. Again, published discussions of economic, managerial, and industrial uses tend to explore the possibilities and implications, rather than to provide clear-cut demonstrations of practical value. On the whole it would, however, be generally agreed that game-theory has made enormous contributions to our orientation with regard to a wide range of very complex problems. The concepts of players, pure and mixed strategies, pay-off matrices, etc., provide a new system of thinking about complicated situations involving conflict and competition. For such reasons game-theory must be regarded as one of the basic approaches of the operational research worker in constructing quantitative models of highly patterned decision-making processes.

We have not attempted above to give an exhaustive account of the various techniques and approaches which are given special emphasis in operational research, but rather to highlight a few of these methods so as to mark the distinction between the kind of mathematical models arising in this discipline as compared with the kind of pure and applied science dealt with in earlier chapters. For further discussion of such topics as flow-graphs, information theory and feedback, critical path analysis, etc., the reader must consult one or more of the various textbooks available.

7.4 THE MULTIDISCIPLINARY APPROACH

Science in general and applied mathematics in particular have always been rather eclectic and empirical in their approach to problems. Any device would do so long as it worked, and if the various devices used at different times were to some extent inconsistent this was of no consequence in the short run, though attempts to achieve integration in the long run were desirable and would probably lead to new advances. For example, within the sphere of mathematical physics corpuscular models and wave models of the propagation of light were used when convenient and quite successfully, though a better foundation for all the phenomena concerned was eventually afforded by the more universally applicable explanantions of wave mechanics. More generally, the prosecution of research in any particular scientific field, such as biology, frequently involves the use of knowledge derived from other fields, such as physics, chemistry, statistics, electronics, engineering, etc. Some subjects are outstanding in the way they impinge on a large number of other disciplines. Genetics, for example, overlaps botany, zoology, cytology, physiology, haematology, molecular biology, microbiology, eugenics, anthropology; forensic medicine, statistics, computer science, etc., to mention only a relatively few items. As a rule, new subjects which begin by developing round an area at the periphery of several well-established subjects have a special role to play, both in advancing their own particular interests as well as in influencing the overlapping disciplines. Biomathematics is itself rather like this. To begin with it has seemed to be rather a specialist study primarily for those who happen to possess advanced mathematical skills and who find it amusing to exercise these in a small corner of the whole biological field. Gradually, however, as indicated in earlier chapters of this book, the use of quantitative methods has come to pervade vast areas of biological science, not only making specific contributions to individual topics of research but also establishing new connexions and relationships between diverse activities by means of an identical underlying methodology.

Multidisciplinary activity has therefore long been characteristic of much scientific work, though the disciplines involved have usually been scientific ones of not too disparate a kind. Operational research carries this process of broadening the field of relevant disciplines a stage further, so that a multidisciplinary approach now becomes a deliberately fostered technique of investigation. It is easy to see how this comes about. Consider, for instance, the historical origins of operational research, particularly the massive developments during World War II referred to in Section 7.2.

These military problems were typical of operational research in that from the start they entailed coping with difficulties arising from the attempt to organize and control a number of widely different elements. On the personnel side there were members of the armed services, some carrying out routine operations with specially designed equipment, some responsible for communications and supply, others concerned with the planning of armed intervention at various levels. Deeply involved as well were various civilian scientists and engineers, together with the relevant industries, whose task was to aid the development and manufacture of weapons, ships, aircraft, tanks, etc. Again, at a high level of direction and planning, various governmental agencies were implicated. To find the best way of improving the conduct of any campaign or action entails not only a painstaking study of the overall strategies and tactics, at a planning level, but necessitates taking into account the physical potentialities and shortcomings of the hardware as well. This might well mean detailed discussions with experts on high explosives, ballistics, aircraft engineering, direction finding, economics, psychology, etc. In short, practically all man–machine systems have a large number of aspects which can be classified under a wide range of scientific, industrial and managerial disciplines.

It thus follows that any successful operational research investigation will almost certainly have to be carried out by a team of people, each with some special knowledge covering only a part of the field (see also Sections 9.3 and 9.5). Thus a team with which the writer was personally concerned, working on the function and design of hospitals, included an architect, a sociologist, a doctor, a nurse, an historian, a retired army officer, and a statistician. Part of the value of such a mixed team is in having all main aspects of the central problem adequately covered, though normally it is not necessary to have experts in every possible subject that may arise. Another part of the value resides in the fact that executives and administrators have precisely this difficulty of trying to organize activities in which they themselves are not experts. The mixed operational research team therefore reproduces this problem in some degree in an environment where an understanding and resolution of the main difficulties are greatly facilitated. If the members of a broadly based team can learn to work together, to communicate with each other, to learn something of each other's technical languages and modes of thought, and to cooperate in finding practical solutions, then they are much more likely to be able successfully to transmit their ideas and proposals to the administrators who must ultimately apply the fruits of research. The multidisciplinary approach is thus an integral part of operational research, and facilitates not only a study and analysis of the basic problem but also the implementation of its solution.

7.5 THE DEMAND FOR MEDICAL CARE

One of the most important aspects of providing medical care is the problem of estimating to a reasonable degree of accuracy the volume of services and accommodation required. Distinctions must, of course, be made between such main divisions as general practice, hospital outpatients, hospital inpatients, and home cares schemes, as well as other local authority services. Classification by medical specialty is also required as a rule, especially for hospital patients.

In particular, the importance to the community of obstetrics, geriatrics, and mental health must not be forgotten. Any one of these fields presents difficulties of enormous complexity. Moreover, the different fields are to some extent overlapping, and a complete analysis of the structure of the 'total system' has yet to be undertaken. The details of organization and structure naturally vary a good deal between one country and another, although the overall aims and objects remain roughly the same.

Let us, by way of illustration, consider the problem of studying the demand for hospital beds in relation to the community served. The pioneer research in this field, usually referred to as 'case-load studies', was done by the Nuffield Provincial Hospitals Trust (1955), and essentially consisted in relating the actual volume of demand for beds to the effective population at risk. Distinctions were made between different specialities, and areas were chosen where general conditions were thought to be fairly good so as to avoid distortions due to serious underprovision.

Previous to the Nuffield work (and indeed all too frequently since) estimates of the numbers of inpatient beds required were largely based on sheer guesswork, after, of course, careful consideration of all the various factors involved. Planners would use figures like 200 general acute beds per 100,000 of population, or 450, or 700, according to the authorities favoured. Different estimates could show wide variation, as much as an order of magnitude in some cases! Thus certain estimates of the number of thoracic beds needed per million varied between 10 and 120. The trouble was that although figures could be arrived at, and sometimes generally agreed, after mature deliberation, there seemed to be no proper rational basis for their derivation. In other words, there was no acceptable underlying mathematical model and no way in which true statistical estimates with standard errors could be obtained. Consequently, it was difficult to know what use could be made of the figures in quantitative discussion, in trying for example to plan for a new town where a number of relevant factors might operate quite differently from the areas covered by previous survey.

The first step was to investigate the populations actually served. Although (in Britain at any rate) medical administrative authorities work in terms of well-defined geographical areas containing known populations with specific groups of hosptials serving them, the actual pattern of hospital admissions is considerably more complicated than this. Thus, patients needing admission may have a certain choice of hospital, which depends partly on their own wishes and partly on their general practitioner's policy. In general, patients tend to go to hospitals that are more rather than less accessible, even if the nearest hospital is in a different administrative region from the one they live in. Investigation shows that a given hospital or closely related group of hospitals is likely to receive nearly 100 per cent of the patients who require admission and who live in the immediate vicinity. As we go further away from the hospital this percentage falls steadily until it is practically zero in the neighbourhood of an adjacent group of hospitals. Midway between two large groups we may find an admission percentage near to 50. Thus the admission rate to a given hospital may vary continuously with distance, irrespective of where administrative boundaries occur.

The effective population served by a chosen group of hospitals can be estimated as follows. We take a list of patients admitted to the group during a whole year, say, and classify the patients (by means of their addresses) according to the geographical areas from which they have come, e.g. urban and rural districts. This, in general, reveals only part of the demand from any particular district, as some patients may have gone to other hospitals. The analysis is therefore repeated at all surrounding groups of hospitals that are likely to attract an appreciable number of patients who might otherwise have gone to the groups chosen for investigation. For the ith geographical district we can now distinguish two groups of patients, a_i who went to the main group under study and b_i who went elsewhere. If census data show that the district contains N_i individuals we can estimate the *effective* population served in that district by the group of hospitals under study as $a_iN_i/(a_i+b_i)$. Summing over all relevant districts, whether inside or outside the official administrative region, then gives the estimated total effective population served, M, as

$$M = \sum_i \frac{a_iN_i}{a_i+b_i}. \tag{7.13}$$

A standard error for M can easily be found, depending to some extent on the assumptions made about sampling behaviour. If the a_i and b_i can be regarded as Poisson variables, as is often quite reasonable, the large-sample variance of M is easily seen to be

$$\text{var}(M) = \sum_i \frac{a_ib_iN_i^2}{(a_i+b_i)^3} \tag{7.14}$$

assuming the N_i to be known more or less exactly.

Applying these formulae to material collected in the Norfolk and Norwich region for 1951, the estimated effective population was found to be 484,600 $\pm$ 800, based on an analysis of 41,000 addresses for a total of some 60 urban and rural districts. The actual population of the official administrative area was approximately 546,600, which exceeds the effective population by nearly 13 per cent. In some cases, the divergence of the two figures might be even greater, depending very much on local conditions.

Having dealt with estimating the population at risk, we next consider measuring the actual demand for inpatient care in the several specialties of the main group of hospitals under study. This is fairly straightforward, since demand can be assessed in terms of admissions during a year (or discharges and deaths) plus any changes that may have occurred in the relevant waiting-lists, where these exist. Providing that the latter are accurate and up-to-date, increases represent additional demand that must be added to the admission figures, while decreases have to be subtracted.

Now if the average demand per day in a given specialty is d patients and the average length of stay is s days, we should expect the average number of beds required to be c, where

$$c = ds. \tag{7.15}$$

Although this is approximately correct, there is an important qualification. Where there is a fairly heavy demand for limited inpatient accommodation, quite possibly with a wating-list building up, we have what is essentially a queueing process. As indicated in Section 7.3, it is vital to ensure that the traffic intensity is less than unity if very long (theoretically infinite) waiting-times are not to build up. This means that in practice we always need slightly more accommodation than the so-called *critical* value calculated from (7.15). The additional amount required may be quite small (see Bailey, 1954), as can be seen from examination of a simple queueing model.

Let us suppose that we have a ward with k beds, and that the length of stay is very variable, having in fact the negative-exponential distribution

$$f(v) = e^{-v/s}dv/s \qquad (0 \leq v < \infty) \tag{7.16}$$

where the average length of stay is s days, as above. Let us also suppose that the daily demand for accommodation has a Poisson distribution with mean d. Then the traffic intensity ρ is given by

$$\rho = \frac{sd}{k} \tag{7.17}$$

and this is also the basic bed-occupancy, apart from further practical administrative difficulties in filling beds. If the process has settled down to an equilibrium state, with $\rho < 1$, it can be shown that the average time w that a patient would have to wait for admission would be

$$w = \frac{s}{(1-\rho)k}\left(1 + \frac{(1-\rho)\{1 - P(k, sd)\}}{p(k, sd)}\right)^{-1} \tag{7.18}$$

where $p(x,m)$ and $P(c,m)$ are the terms and sums of terms of a Poisson series, conveniently tabulated by Molina (1942).

Consider, for example, a ward in which the demand is exactly one patient per day, i.e. $d = 1$, and the average length of stay $s = 20$ days. We might have 25 beds in all, but allowing for administrative difficulties in arranging for beds to be filled can only keep 22 occupied when there is a waiting-list. For our purposes $k = 22$, and $\rho = 0{\cdot}91$. Applying (7.18) we easily find that $w = 5{\cdot}7$ days, which in many cases would be regarded as a fairly short waiting-time (except for emergencies which have to be catered for separately). In general, we need to supply only one or two beds over and above the critical value.

In estimating the critical number c from actual data, we need a standard error. If d is estimated from a total of n new arrivals (some being admitted, some going on to the waiting-list) in a particular period of time, and if s is calculated from n' observations, the large-sample variance of c is

$$\text{var}(c) = c^2\left(\frac{1}{n} + \frac{1}{n'}\right). \tag{7.19}$$

Finally, the estimated critical number of beds needed per head of population is given by

$$R = c/M \tag{7.20}$$

where

$$\text{var}(R) = R^2\left\{\frac{\text{var}(c)}{c^2} + \frac{\text{var}(M)}{M^2}\right\} \tag{7.21}$$

combining the results in (7.14) and (7.19). We now have a statistically well-founded estimate of the quantity R together with an appropriate standard error, which means that further discussion can be put on a much sounder numerical basis. In planning the hospital facilities for a new town, for instance, we could apply known values of R for different specialties to the new proposed population of N_0, say, estimating the relevant critical number as RN_0.

Of course, in such an application we should have to be reasonably sure that the general hospital and population conditions in the new town were not too dissimilar from those from which R was originally derived. Discussion here of the various complications of this problem would take us too far afield. Suffice it to say that we may be in a position to use some obvious modifications of the above formulae. For example, we may accept that the average demand under new conditions will be roughly the same as elsewhere, but that the medical staff will wish to work with a different value of the average length of stay s. We then apply formulae such as (7.15) or (7.20) using the proposed value of s instead of the value actually observed in the material surveyed.

One common criticism of the above approach is that we are, at best, measuring only current *requests* for inpatient care as opposed to real *need*. This is perfectly true, but until a better assessment of actual need is possible an administrator can do no more than plan to meet the requests which are expected to arise. Again, some research (e.g. Forsyth and Logan, 1960) has tended to demonstrate the converse difficulty of 'clinically unnecessary admissions', which inflate the observed number of requests. This usually refers to admissions that have been strongly influenced by social and economic factors.

If adequate home-care schemes are introduced, the demand for inpatient accommodation can naturally be reduced, but not otherwise.

The basic arithmetic of supply and demand, as described above, is still valid even when the complications just mentioned are taken into account. We must, however, decide, for the hospital or community about which we wish to make predictions, just what the relevant values of d and s are likely to be. Given these values, we can calculate the corresponding critical number c, and hence the practical consequences of alternative proposed degrees of provision.

In some circumstances, the primary planning problem can be met more easily. Suppose, for example, that an existing hospital is in some way to be rebuilt, remodelled, or reorganized, and a new assessment of the actual demand is called for with a view to estimating the accommodation really required. If, subject to certain minor modifications and adjustments, we can take the existing demand as given, there is no point in undertaking an extensive case-load type of study to determine the effective population served. We merely want to know how many beds are needed to deal with so many patients. Such a situation arose in 1960 in connexion with the replanning of the United Oxford Hospitals. The population served by this teaching group was known to be extremely diffused and to cover a large geographical area (the records at some 400 hospitals would have had to be surveyed if a case-load type of investigation were to be carried out). Moreover, it was not clear with what other areas the Oxford hospital group could legitimately be compared. Again, some specialties had marked trends of demand, and there was also much discussion about the clinically desirable average lengths of stay.

Consider, for instance, the summary of the demand for general surgery over the previous five years shown in Table 7.1. Thus the total actual demand in 1959 was 5820, quite irrespective of the effective population from which this demand arose. With an average length of stay of 9·8 days and an average turnover interval of 2·2 days (i.e. the average time a bed is vacant between successive patients due to administrative difficulties in filling it) the critical number of beds required would be

$$c = \frac{5820}{365} \times (9{\cdot}8 + 2{\cdot}2) \text{ beds} = 191 \text{ beds.}$$

Table 7.1 Demand for general surgery at the United Oxford Hospitals

Year	No. of discharges	No. on waiting list at 31 Dec.	Change in waiting list	Total demand	Average stay in days
1955	5176	1176	+ 193	5369	9·2
1956	5256	887	− 289	4967	9·2
1957	5260	1163	+ 276	5536	8·9
1958	5459	1413	+ 250	5709	9·8
1959	5583	1650	+ 237	5820	9·8

Since only 183 were on average available, it is clear that the traffic intensity at the time in question was somewhat greater than unity. A steady increase in the waiting-list would therefore have occurred, even if there had not been a marked trend of increasing total demand.

It is obvious that the average length of stay plays a fairly dominant role in all these calculations. The actual stay of individual patients can, of course, be highly variable, but the average depends partly on the range of clinical conditions involved, partly on medical practice, and partly on what accommodation is available. It is therefore to some extent capable of being changed or modified, according to circumstances. Thus surgeons or physicians may regard an actual average length of stay as inadequate, but arising from heavy pressure on beds. Alternatively, a reduction might be contemplated if appropriate home care facilities were available. At the United Oxford Hospitals the surgeons concerned felt that an average stay of 10 days was too short. Most preferred as much as 15 days; though some thought 12 would be sufficient. Whatever figure one decides on, the corresponding critical bed complement is easily calculated. Thus with a 12 day stay and a 2 day turnover interval, the 1959 demand would entail a critical value of

$$c = \frac{5820}{365} \times (12 + 2) \text{ beds} = 223 \text{ beds}$$

a value appreciably greater than that actually available at the time.

Such calculations do not, of course, provide precise unqualified estimates of what is needed. But, given certain explicit assumptions, they allow one to compute their automatic consequences. Planning is accordingly made much easier, as the implications of various alternative schemes can be worked out, and this supplies a numerical basis for further argument and discussion. Let us consider again the data of Table 12.1. Suppose that the trend of about 2 per cent annual increase in demand were to continue. How could any satisfactory solution be found? One possibility, out of many, would be to aim at providing the 15 days average stay that most surgeons said was necessary, for which 272 would be the relevant critical number of beds; then to operate at a level of 12 days for a year. This would deal with a volume of about

$$\frac{272 \times 365}{(12 + 2)} = 7100 \text{ patients}$$

in a single year, and would practically wipe out the waiting-list. Thereafter the full 15 day average could be used, but with slight progressive reductions, if the trend continued, to prevent re-establishment of a long queue.

The above ideas are, of course, only an elementary introduction to the principles involved (see Bailey, 1962, for a more detailed discussion and further references), but they do illustrate the way in which a simple mathematical model can be used to sort out in a realistic manner many of the fundamental problems of deciding the general scale of inpatient accommodation.

7.6 THE DESIGN OF HOSPITALS

The approach of the previous section was primarily concerned with establishing the general scale of inpatient provision in terms of the number of beds required in different specialties. Let us now suppose that this has been done in relation to some new hospital that is to be constructed. How, precisely, should the building be planned so as to provide the highly complex pattern of clinical services required in a functionally efficient way? To a great extent this is basically an architectural problem. In principle, the client says what he aims to achieve and the architect supplies one or more possible solutions, having due regard to the requirements of aesthetics, economics, engineering practice, and functional performance. In practice, a hospital committee may inspect the elevations of a number of plans by different architects and simply choose the one they think looks best. No one doubts that the clients know what they want in general. The difficulty is to translate the overall aim of efficiently providing medical care into terms that can be related in detail to specific architectural designs.

The day is probably a long way off when it will be possible to compute mathematically a complete design for meeting specified requirements in the most efficient way. What can be done, however, is to calculate some of the consequences of any given design so that the quantitative performances of alternative proposals can be compared. In certain respects, therefore, we can arm ourselves with numerical criteria where previously there was nothing but sheer guesswork. Such an attitude of mind should enable hospitals to state their basic requirements more clearly and architects to know whether any tentative plans they have in mind are likely to be acceptable. It is difficult to foresee how things may develop, but it is clear that all this is very much a first stage in the establishment of much more quantitative and scientific ways of thinking about building design.

Let us first look at a number of problems connected with individual wards, as these can to some extent be considered independently of the general design of the whole hospital. Again, most of the basic work was originally undertaken by the Nuffield Provincial Hospitals Trust (1955), though for a more statistical discussion Bailey (1956) should also be consulted.

One of the most important aspects of ward layout is the need to have a relatively compact design. If a long ward is used, with the various ancillary rooms all located at the end, nursing staff are likely to be subjected to an unnecessary amount of walking. The relative merits of different designs may not be immediately obvious to simple inspection, since the pattern of a nurse's movements between beds, duty-room, sluice-room, etc., also has an important influence on the total walking distance. One way of measuring this activity, which has never really been tried, is to establish an average standard pattern of movement between all the basic elements of a ward unit by carrying out surveys of staff movements at a suitably chosen sample of hospitals. The standard pattern, for any given specialty, would therefore specify the average number of journeys per day or per shift between, say, the duty-room and the bed area. It might also be necessary to describe more precisely the pattern of movement within a ward between the different beds. Anyway, whatever method was adopted of arriving at a suitable standard, the latter could be applied to a series of alternative ward layouts and would provide a measure of relative compactness. This index would not, of course, have an overriding influence, but would be just one of the numerical parameters to be taken into account. Other things being equal, we should choose the most compact design which entailed the shortest walking distance per spell of duty.

There are, however, limits to the compactness which is practical. For example, the need to avoid overcrowding to reduce the risks of cross-infection, and the importance of there being sufficient space around beds for nurses to perform their duties properly, both entail some decision about how far apart beds should be. A quantitative measure of the connexion between overcrowding and cross-infection has still to be developed, but the nursing requirement is more easily settled. In the Nuffield studies referred to, cinematograph recordings were made of various nursing procedures actually in progress. The floor surrounding the patient's bed was marked out in one-foot squares, and a clock was installed at the end of the bed. It was then a simple matter to construct statistical frequency distributions of the relative proportions of time that different amounts of space were required. Thus in bedmaking 86 per cent of a nurse's time was spent inside a line 2 feet from the bed, while 98 per cent was within 2½ feet. For the more complicated procedures of giving an intravenous infusion or helping a patient into a wheel-chair more space was required, but this never exceeded the 4-foot mark. In certain special cases additional space was necessary. The arranging of an oxygen tent over a patient in bed involved two nurses and two orderlies. About 28 per cent of the whole operation was beyond the 4-foot line, but the total time in absolute terms was only about one minute. Detailed statistical analysis was clearly unnecessary. For practical purposes a 4-foot separation between beds would give a comfortably sufficient amount of space. Only rarely would a nurse find conditions slightly cramped. If beds are taken to be 3 feet wide, it is evident that about 7 feet between bed centres should be adequate.

A more general feature of ward design is the provision of facilities such as lavatories, bathrooms and day spaces. The extent to which these are required depends very much on the numbers of ambulant and partially ambulant patients, and this in turn depends on specialty as well as medical policy. Obviously a highly variable quantity is involved with considerable day-to-day fluctuations, but some numerical information on its statistical nature was

Table 7.2 Percentages of surgical patients who were ambulant under different policies

Type of policy	Bedfast	Partially ambulant	Fully ambulant
Traditional	67	17	16
Actually observed	50	20	30
Early ambulation	16	35	49

obtained as follows. A doctor carried out a survey over a period of time on patients in different types of wards and recorded not only the actual numbers of bedfast, partially ambulant and fully ambulant individuals, but also the numbers that might have been in these three categories if (i) a more traditional policy had been adopted, and (ii) if there had been some bias towards a regime of early ambulation. Naturally a good deal of personal judgement had to be used, but the numerical comparisons for surgical patients shown in Table 7.2 are most illuminating.

The variation between the percentage of bedfast patients under a traditional policy and the percentage expected with early ambulation is quite striking, and the consequences for ward design could be considerable. In order to apply such figures, calculations would have to be made for a given ward of the actual numbers of patients expected to use the various facilities. To do this properly up-to-date survey material would be required, as well as information about the type of ambulation policy to be operated. Obviously there would be uncertainties and difficulties, but the point is that such numerical reference material would greatly facilitate discussion between architects and doctors about what was required.

Another aspect that affects basic ward design is the extent to which provision should be made for a certain number of beds to be installed in separate cubicles, as opposed to the main ward. There are many practical reasons for wanting a certain amount of accommodation of this type, some medical such as a patient's being infectious or unduly susceptible to cross-infection, some social such as his being noisy or having distressing symptoms. Again, there are bound to be large fluctuations in demand, but the problem of optimum provision is an important one. It so happens that a more detailed mathematical analysis is possible and indeed necessary to bring out the salient features of the problem. Moreover, the analysis illustrates a general difficulty that frequently arises, namely the conflict between opposing requirements. In a general way it can be seen that if all beds were installed in single rooms then there would be no demand for separate accommodation that went unsatisfied (except in the rare event that the whole ward was overfilled with emergencies needing to be isolated). But a great deal of the time the special accommodation would not be used for its proper purpose. This might be adjudged uneconomic, and could easily be inconvenient for nursing staff.

On the other hand, if only a very small number of single rooms were provided they would nearly always be properly filled, but would leave much of the demand unmet. We clearly need to strike some kind of balance.

For a proposed ward we could certainly determine from a survey of the type of patients expected (e.g. in other hospitals elsewhere) what the probability would be of a given patient requiring separate accommodation on any one day. Hence we could say, for a particular ward size, what average number of patients would need a single room. However, because of the large statistical fluctuations occurring in small wards this average number is not necessarily optimal. It is easier to see what would be achieved by any arbitrary degree of provision if we clearly distinguished between (i) *the efficiency of provision*, i.e. the percentage of actual demand which can be met, and (ii) *the efficiency of use*, i.e. the percentage of time that the provision would be used for its proper purpose. For example, it was found in one survey of medical patients that the chance of an individual requiring a separate room on any one day was about 0·3. Hence, the average demand for a 16-bed ward would be $0{\cdot}3 \times 16 = 4{\cdot}8$ beds, say about 5 beds. But what would this amount of provision actually achieve in practice?

A fairly simple mathematical model is as follows. Suppose that the chance of a given patient requiring a single room on a given day is p. Then for a ward with n beds the chance of there being r patients needing such accommodation is the binomial probability p_r given by

$$p_r = \binom{n}{r} p^r (1-p)^{n-r} \tag{7.22}$$

the average demand being np. Suppose we now imagine that an arbitrary number a of single rooms has been provides. The average number S of patients in such rooms who really require to be isolated is thus

$$S = \sum_{r=0}^{a} r p_r + \sum_{r=a+1}^{n} a p_r. \tag{7.23}$$

The first term in (7.23) represents the contribution from days when the demand can be fully satisfied, while the second term relates to occasions when demand outstrips supply. If the efficiency of provision is E_{p} and the efficiency of use is E_{u}, it is easy to see that

$$E_{\mathrm{p}} = S/np, \quad E_{\mathrm{u}} = S/a. \tag{7.24}$$

Simple calculations based on (7.23) and (7.24) give the results shown in Table 7.3.

If we actually provide 5 separately accommodated beds, we shall satisfy 87 per cent of the demand and the beds would be used for their proper purpose 83 per cent of the time. It might easily be decided that this was in fact

Table 7.3 Efficiencies of provision and use in a 16-bed medical ward ($p = 0{\cdot}3$)

No. of single rooms	Efficiency of provision	Efficiency of use
2	41	99
3	60	96
4	76	91
5	87	83
6	94	75

an optimal balance, though the efficiency of provision could, for example, be increased to 94 per cent at the expense of reducing the efficiency of use to 75 per cent. If we provide less accommodation, on the other hand, the balance swings the other way. In general, if the number of separately accommodated rooms is approximately equal to the average demand (4·8 beds in the above example) a fairly optimal balance between the conflicting requirements will be obtained in large wards. But with small units it is difficult to maintain high values of both efficiencies simultaneously. The use of computed figures like those in Table 7.3 then becomes correspondingly more valuable. Presenting the results in this form does not enable the mathematician to tell the medical staff what they want or the architect what plan he should choose, but it does ensure that everyone has a more accurate picture of what alternative designs imply. Joint discussion between clinicians and architect can then proceed on a more rational basis.

There are, of course, many other aspects of ward layout which can be subjected to quantitative analysis with a view to assisting the more efficient formulation of new designs. The same is true of most other hospital facilities, and operational research investigations have been applied to operating theatres, X-ray departments, outpatient clinics, etc. Apart from questions of improving the functional efficiency of individual sections of a hospital, there is also the question of their interrelation. Consider, for example, the internal traffic of a hospital involving at its simplest the movement of human beings and various material objects. In general, areas with a large volume of mutual traffic should be sited fairly close together, while those with less traffic can further apart. This principle may, however, have to be modified by matters of urgency or administrative convenience. Broadly speaking, there are two main aspects. First, there is the *total flow* of traffic between any two areas, which we hope to reduce to a minimum so as to cut down walking distances for patients and staff, and also the amount of noise and dirt. Secondly, there is the *traffic density* which will vary with time and place. This too must be reduced as far as possible, at least to the point of avoiding local congestion and bottlenecks.

The main problems usually arise in connection with outpatients, visitors, and passenger lifts, and it is often sufficient to scrutinize proposed hospital plans with these chief items in mind. Detailed numerical estimates of traffic flow and density can be obtained by elaborate and costly surveys of existing institutions, but the figures may not be readily applicable to a new design. It is probably preferable to attempt to construct *a priori* predictions along the following lines. The overall volume of outpatient attendances in any specialty can be derived from the type of investigation described in Section 7.5 for inpatients. The detailed pattern of outpatient traffic depends partly on design and partly on the clinic program, so that the proposed *organization* of outpatient services must be taken into account as well.

Traffic due to visitors is often in the region of 2 per inpatient. The consequences for traffic flow then depend on the timing of visiting hours and the siting of stairways and lifts. Again, the actual organization of this activity has an important influence on whether serious traffic problems will arise or not.

The problem of lift traffic is more difficult to deal with, and installations may frequently be observed to be inadequate in practice. Lift accommodation is often based on commercial rule-of-thumb methods having no rational basis, and unlikely in any event to apply to hospitals. It is easy to see that lifts involve some quite complicated queueing processes. Whether detailed mathematical investigation is worthwhile is a matter for speculation. It might well be useful to be able to calculate in advance the consequences of different types of proposed installations.

7.7 OUTPATIENT APPOINTMENT SYSTEMS

Although part of the problem of achieving functional efficiency in hospitals has to do with general design, layout and scale of provision, a vitally important aspect is concerned with the organization of the available facilities. As we saw at the end of the last section, the avoidance of serious internal traffic problems may depend as much on the planning of clinic sessions and visiting hours as on the physical siting of wards and outpatient departments. The general sequencing of activities within a limited sector can often be studied in relative isolation. A good example of this is the running of an outpatient department, which requires a fair amount of administrative skill quite apart from the specialized clinical and nursing services involved.

There have been constant complaints for a great many years about the excessive periods of waiting endured by large numbers of outpatients. A Nuffield Provincial Hospitals trust (1955) survey carried out in 1952 revealed an average waiting-time of 56 minutes between appointment and actual consultation, taken over 34 different clinics. At one London teaching hospital surgical patients waited on average 1 hour 40 minutes, 17 per cent having waits of over 2½ hours and about 2 per cent waiting as long as 3½ hours. A subsequent more extensive survey by the Nuffield Provincial Hospitals Trust (1965) completed in 1964 showed an overall average of 25 minutes, which suggests some improvement although the two samples are not strictly comparable. In the later survey taken as a whole 34 per cent of patients had to wait over half an hour, and 11 per cent more than one hour. There was considerable variation between hospitals. Some were appreciably better than the overall figures suggest, but others were much worse.

It is found in practice that certain outpatient departments have comparatively streamlined appointment systems with quite short waiting-times, while others appear to be rather inefficient, all patients being told to arrive at the same time and having, therefore, long periods of waiting. There is still a good deal of controversy about this subject: while some consultants believe that a substantial reduction of waiting-time is possible, others insist that a lot of waiting is inevitable. It may happen that a particular clinic has arrived at a generally satisfactory working arrangement almost by chance, but once established it tends to persist. Difficulties arise where conditions are bad. It is then often thought that the whole situation is so complicated and the intrinsic factors so variable that no appreciable improvement could be obtained by any means, although it is usually accepted that excessive waiting is highly undesirable on many grounds, clinical, economic, humanitarian, etc.

The writer believes that the case for having a suitably designed appointment system is proved beyond all reasonable doubt and that this view is supported by common sense, observation, and theory. By common sense, since patients' waiting could clearly be reduced quite substantially by having a sufficiently large interval between successive appointments. This might, of course, lead to the consultant having to wait for patients. It would then be a problem of balancing conflicting requirements, as with the single-room situation in the last section. By observation, since some clinics actually do succeed, and since many commercial enterprises also manage to prevent their customers having to suffer excessive waiting. And by theory, since a queueing process type of investigation can be carried out, using a model exhibiting the main features of the whole set-up, which demonstrates that satisfactory optimal appointment schemes are perfectly feasible.

Let us examine the role played by such theoretical studies. In order to have a proper rational basis for discussion, we need a mathematical model of the progress of patients through an outpatient clinic. This model should in the first instance be fairly simple, exhibiting most if not all of the salient features of a real clinic. It should certainly make allowance for the highly variable length of individual consultations, and should permit the comparison of different types of appointment system, especially with regard to patients' waiting-time and possible consultant's waiting-time.

If we think of the simplest arrangement of a consultant in effect providing a service to a succession of customers, it is clear that we have some form of queueing process (see Section 7.3), which should be susceptible of standard mathematical analysis. There is, however, a difference here from some applications. In Section 7.5 we made use of a certain kind of queueing process to represent the way in which a ward deals with the demand for inpatient care. For practical purposes, we merely wanted to calculate quantities like the average waiting-time before a given patient could be admitted, assuming that the process had settled down to an equilibrium form of behaviour. Such averages, taken over ideally infinite periods of time, are often fairly easy to calculate. But a typical clinic session is only of limited extent, and the mathematical theory of queueing processes over finite periods of time is highly complicated even for the simplest models. Since the outpatient process is not in general even a Markov process, it is more easily investigated by simulation or Monte Carlo methods (see, for example, Sections 4.5 and 7.3).

Let us envisage a clinic session with 25 patients, and suppose that the individual consultation times are highly variable, the frequency distribution being essentially a χ^2 with 5 degrees of freedom (apart from a change of scale), which is very similar to many observed distributions. To fix ideas it is convenient to assume an *average* consultation time of 5 minutes. In the current context we use the term 'consultation time' to cover the whole period during which a patient is claiming the consultant's time or attention. It thus includes history-taking, clinical examination, the giving of advice, and a certain amount of undressing, while in teaching hospitals teaching time will be included as well. The whole clinic session, so far as the consultant is concerned, will last on average about 2 hours 5 minutes, unless he is himself kept waiting for patients for any appreciable time during the clinic. We suppose that the patients are seen in succession, and that they arrive according to some predetermined rule. This rule constitutes the appointment system, and a number of alternatives can be studied and compared.

Using an appropriate statistical technique for selecting random variates from the χ^2 distribution, we can build up an artificial list of 25 typical consultation times, which can be regarded as representing a simulated group of patients attending a single clinic. An investigation by the writer (Bailey, 1952) employed a series of 50 such 'mock' clinics, and these were used to test the consequences of different assumptions about the manner of arrival. Suppose, for instance, we imagine the consultant to start work with an initial block of 5 patients, the remaining 20 being timed to arrive at 5-minute intervals. It is clearly a matter of simple though tedious arithmetic to work out how long each patient would have to wait for his consultation, and conversely how long the doctor might have to wait for a patient after the completion of the previous consultation. In the original study ordinary pencil-and-paper methods were used, but in modern, more extensive, investigations automatic computation would be standard.

The effects were examined of varying factors like the time-interval between arrivals or the size of the initial batch of patients. In general, it seemed best to make the inter-arrival time equal to the average consultation time, but the choice of initial block size was crucial. This is illustrated in Table 7.4, which shows the average patient's waiting-time, together with the correspond-

Table 7.4 Patient's average waiting-time and consultant's free-time per clinic according to size of initial batch of patients, for an average consultation time of 5 min and appointments at 5 min intervals

Initial batch of patients	Patient's average waiting-time (min)	Consultant's average free-time per clinic (min)
1	7 (18)*	9
2	9 (20)	6
3	12 (23)	3
4	16 (27)	2
5	20 (31)	1
6	24 (35)	$\frac{1}{2}$

* One patient in eight waits longer than the number of minutes shown in brackets.

ing average consultant's total waiting-time per clinic (or free-time), according to variations in the size of the initial batch of patients.

This table shows very well the conflict of opposing requirements. The patient's average waiting-time can be reduced only at the expense of the total time waited by the consultant per clinic, and conversely. Of course, what frequently happens in practice is that the consultant heavily over-insures against the possibility of his ever having to wait by starting off with a very large batch of patients. Indeed, Table 7.4 shows that his free-time would total only about $\frac{1}{2}$ min per clinic on average with an initial batch of 6 patients, although the average patient's waiting-time would be as high as 24 min, with the patient's total waiting per clinic being $24 \times 25 = 600$ min. A much more reasonable balance would seem to be achieved by having only 2 patients present initially, when the patient's average waiting-time is 9 min and the consultant's free-time is 6 min. And it must be remembered that even the latter is likely to be largely eliminated in practice by some patients arriving early and by the consultant having to do a certain amount of unavoidable clerical work such as reading, making notes, or using the telephone.

The effect of changing the shape of the distribution of consultation time was also studied, but the general results were fairly insensitive to variations in the number of degrees of freedom between 4 and 7.

It will be noticed that in the arrangement recommended above the traffic intensity is actually unity. If we were dealing with an unending queueing process this would lead to trouble, but with relatively short processes in finite time there is no critical threshold phenomenon.

The simulated clinics described above can also be analysed in other ways, to give for instance some idea of the frequency distributions of the different variables. Thus Table 7.4 gives figures relating to the tail of the patient's waiting-time distribution. Again, it may be useful to look at the distribution of the total number of patients waiting at any one time. Thus it may be shown, for the system recommended above with 2 initial patients, that 22 per cent of the time there would be no patients waiting, while 86 per cent of the time there would be 3 or fewer. Such figures are obviously relevant to the size of waiting space required and may well affect the architectural planning. It would, of course, be necessary to allow for the effects of patients arriving late or early, as well as the presence of escorts.

Although we have discussed at some length the properties of a rather simple clinic queueing model, the general implications are clear. Patients' appointment times must be staggered so as to achieve a reasonable balance with the consultant's free-time. An easy way to do this is to determine the appropriate average consultation time from a fairly large sample of patients and make the appointment interval equal to it (probably rounding off calculated appointment times to the nearest 5 min).

Many, if not most, real clinics are of course more complicated than the model. There may be more than one doctor, e.g. a consultant plus one or more registrars. If we cannot distinguish separate consultations we must treat the group of doctors as a single unit, and ensure that the overall rate at which consultation is supplied is roughly equal to the rate at which demand arrives. Again, some patients may have to go for X-ray, followed by a second consultation. In calculating the average consultation time per patient all the consultation times should be included but each patient counted only once. Further complications arise with some patients being late or early, or defaulting altogether. As a rule, the practical implications of such factors can be calculated quite easily, at least to a sufficient degree of approximation, and due allowance made for them. If there were grave doubts about the likely consequences of some special factor, an appropriately designed simulation study could easily be undertaken. But in general the most important problems are not theoretical, but practical, namely to persuade administrators, both lay and medical, that reasonably efficient appointment systems are, in fact, simple both to design and to operate.

CHAPTER 8

Systems Analysis and System Dynamics

8.1 SYSTEMS ANALYSIS

We suggested at the beginning of Chapter 7 that the term *operational research* could be, and often is, employed to cover the whole area of a scientific approach to the problems facing executive and administrative authorities, and that it could thus include *systems analysis* and *systems engineering* as well. This is all really a matter of convention and convenience. In the last chapter we rather emphasized a commonsense, down-to-earth approach to organizational problems. A practical and pragmatic attitude, coupled with an insistence on operational aspects, is the essence of all operational research undertakings—although, of course, there are, as we have seen many highly specialized techniques that can be effectively used in appropriate circumstances.

However, the very success of operational research has over the years led to the development of (*a*) increasingly sophisticated technical methods of investigation, and (*b*) a more conscious systematization of the operational research approach itself. We shall discuss some of the newer possibilities, especially system dynamics, later in this chapter. As to (*b*), there has been a certain tendency to make more use of the term *systems analysis* in recent years, giving greater weight both to the logical structuring of the problems to be investigated and to the investigatory approach as such.

Actually, some 20 years ago, Churchman, Ackoff, and Arnoff (1957) were distinguishing the following major phases of an operational research project:

1. Formulating the problem.
2. Constructing a mathematical model to represent the system under study.
3. Deriving a solution from the model.
4. Testing the model and the solution derived from it.
5. Establishing controls over the solution.
6. Putting the solutions to work: implementation.

Given our previous discussions, most of the above phases are self-explanatory as written. It may, however, be necessary to mention that No. 5 refers to the need to check that the system has not changed in some substantative way, e.g. in structure or parametric values, between the time of solution and the time of implementation, appropriate adjustments being required if this has occurred.

The rather precise sequencing of the operations indicated by the six phases above can be deceptively simple. In many commercial and industrial applications, a substantial degree of quantification exists from the start. Problems easily present themselves in terms of maximizing profit, output, efficiency, etc. Models of now classical type can be developed to deal with inventory situations, the allocation of resources, questions of queueing and traffic delays, the sequencing of assembly lines, the replacement of equipment, and so on, all incorporating clearly defined dynamic processes with economic price-tags on each component. In such applications the operational research approach is essentially straightforward in broad structure, although individual technical problems of considerable difficulty may have to be faced and solved at various points.

Some of the simpler problems in the health field also fall into the same class. We have already examined in the previous chapter certain questions affecting the demand for medical care, the design of hospitals, and the construction of appointment systems. These were fairly typical operational research illustrations. Even though the demands of the six phases were not entirely met in all of the illustrations, it would not be difficult to design more rigorous studies within the prescribed framework if one were undertaking fresh investigations of the same type.

Difficulties begin to arise when one tries, more ambitiously, to handle more complicated systems. Additional complexity may be introduced in order to add a more convincing touch of realism to an oversimplified model, but it then becomes much harder not only to deal with the resultant mathematical analysis but also to specify what overall system of interrelated processes we ought to be modelling, or even to say what broad policy questions we are hoping to answer in the first place. This is what many people have in mind when they talk about the need for a systems analytic approach.

In a more general situation it may therefore be better to start with specifying the overall objective, such as the harnessing of atomic energy or putting men on the moon. The ultimate success of a scientific approach to achieving the latter objectives is now a matter of history. Many of us hope that other problems of high social priority will similarly yield to a concerted multidisciplinary attack, e.g. in the areas of health, education, pollution, crime, transportation, etc.

Once there is a measure of agreement on the ultimate objective, one can begin to move forward. Frequently it helps next to define a number of more specific sub-objectives, but it is usually desirable first to make a broad situational review, or *situational analysis*, of the existing state of affairs. A good deal of judgement is required at this stage as to the degree of detail into which it is necessary to go. Obviously, all major issues and concerns must be included, but basic technical minutiae and routine records would be excluded. Thus in looking at transportation problems on a national level, we can suppose that there is an overall objective calling for 'general improvement', since everyone is more or less directly aware that there is a large number of urgent practical problems and difficulties. We next need to know, in general terms, the overall traffic patterns for road, rail, air, and sea transport, the individual traffic densities in terms of both goods and persons, the operating costs, the risks involved to human life, the dangers of pollution, etc. A situational analysis must, accordingly, describe in round terms all of these things *as they are*. That is, one does not, at this stage, embark on any new major projects of research or data collection, but simply reviews the existing situation, with all its inefficiencies, duplications of effort, absences of vital data, obviously counter-productive activities, and so on.

Such a discussion is at this stage still in general commonsense language. It is practically equivalent to a popular review of the socially visible problems. This situational analysis is then the basis for a more critical discussion. It may readily appear that some more systematic way is needed of satisfactorily codifying and relating the multifarious activities of a vast and complicated mass of interrelated processes. A *conceptual framework* may be needed to provide an agreed basic set of notions, within which a simplified discussion of the major issues can be carried on.

The conceptual framework is therefore intended to clarify thought and discussion about the whole subject, identify gaps in knowledge and inadequate coverage of specific topics, point to inefficient activities, help develop cost–benefit analyses, review decision-making problems, etc. We shall usually find it convenient to distinguish between the technical subject itself, e.g. the engineering potentialities of different kinds of transport in the example above, and the broad administrative and organizational superstructures designed to facilitate general planning and management.

Gradually a pattern begins to emerge. We see many related, though distinguishable, activities, all connected together by certain kinds of interaction. A block diagram or flow chart may help understanding and discussion at this stage. This will probably constitute an information systems model or communication model, of the type referred to in Section 5.3.

We may now be in a position to formulate more specific problems, usually of a decision type. For example, what would happen if we arbitrarily changed the road traffic flows in a given city, by closing streets, changing the directions of flow, altering the time sequencing of traffic lights, building new multistorey car parks, constructing new bypass highways, etc? It may now be possible to make a quantitative description of this more limited system, or perhaps a major part of it, which is in effect a mathematical model of the system or subsystem being studied.

At this point we are very close to the Churchman–Ackoff–Arnoff scheme. The model must be developed, tested, modified, etc., all of which entails considerable mathematical investigation. Solutions of the basic equations finally chosen to represent the system must be obtained and the practical interpretation clearly defined. For practical purposes, this usually means that alternative strategies of control must be identified and the consequences of each described, using appropriate methods of prediction. We have already discussed this aspect of modelling at some length in Chapter 5. Finally, the operational implications have to be implemented, and then evaluated in actual practice.

Bringing together the phases just described therefore yields a modified scheme entailing:

1. Definition of overall objectives.
2. Carrying out a situational analysis.
3. Constructing a conceptual framework.
4. Identifying specific decision problems.
5. Building an appropriate model.
6. Developing, testing, and modifying the model.
7. Specifying alternative control strategies and their implications.
8. Practical implementation of the chosen strategy.
9. Evaluation of the chosen strategy.

Naturally, this is not a rigid framework for all systems analysis investigations: modifications may have to be made to fit different circumstances. But it does bring out the natural progression from the first intuitive feeling for a major, qualitative, human or social concern; through the logical, quantitative specification of a more or less abstract systems formulation; to the final practical evaluation of the strategies actually implemented.

Although the present systems analysis format considerably overlaps the sphere of operational research discussed in the previous chapter, it does have a rather wider domain of application, being better adapted to deal with a complex interacting system of broader and more qualitative issues. In recent years this approach has been applied to an enormous range of complex problems, as can be seen from the extensive literature that has developed. It must be admitted, however, that from time to time there have been and still are serious criticisms of the degree of success achieved. When a fantastically complex problem has a single, clear, agreed objective, like putting a man on the moon, it seems somehow easier to make progress and evaluate results. By comparison, more urgent social problems, such as controlling pollution or eliminating malaria, are much more difficult even to formulate adequately in a realistic way, let alone to find feasible solutions for.

As a result, trenchant criticism of ill-conceived attempts to solve major social problems from a systems analysis standpoint are not hard to find. Take, for instance, the pungent review by Ida Hoos (1969) of systems analysis applied to social policy. She draws attention to the uncritical transfer of successful aerospace systems analysis techniques to the public sector in California in the early 1960s. A number of big, lucrative, but ineffectual, systems analyses were carried out, dealing with criminal justice information, land-use data, solid waste management, and the welfare system. The glamour of 'space magic' was simply being evoked for purely commercial reasons to justify the application of costly 'technological fixes' to fields where they were not appropriate. These criticisms are salutary, but are not destructive of the approach we have been advocating. Complex problems inevitably need special investigatory techniques to facilitate their handling, and a properly tailored systems analysis can be very effective. Nevertheless it is, of course, completely right to castigate fraudulent applications when they occur.

As a final comment on the scope of systems analysis we may consider the partial answer given by Howard Raiffa, when he was Director of the International Institute for Applied Systems Analysis, in the Institute's document *Background Information* published in 1974: 'Our Institute interprets ASA [Applied Systems Analysis] not as a technique but as an embracing rational approach to the resolution of complex problems. ASA is a framework of thought designed to help decision-makers choose the desirable (or in some cases a "best") course of action.'

8.2 HEALTH APPLICATIONS

In the previous section we dealt largely with the general principles of systems analysis, conceived so as to be applicable to a very broad class of highly complex problems, involving definitions of objectives, elaboration of appropriate frameworks of usable concepts, etc., in addition to the more obvious technical components. We also touched on some of the more readily made applications to commerce and industry. We must now go on to see what potentialities there are for the health field itself. In view of the strictures mentioned at the end of the previous section regarding applications to the more difficult areas involving social policies, questions arise immediately as to whether anything useful can be done, over and above the relatively straightforward operational research type of activity illustrated in Chapter 7.

One cannot deny that there is much misgiving in high places about the formal systems analysis approach that starts with a logical clarification of concepts and proceeds in a quantitative direction, though possibly stopping short of a full dynamic mathematical model. Some writers have also claimed that systems analysis has in fact been a more or less total failure in regional health planning. Thus Shuman, Wolfe, and Speas (1974) have forcibly expressed this point of view (using the term 'operations research' in the sense in which we are here employing 'systems analysis'). They suggested a variety of causes, such as that no one had really tried to apply the approach to the work of current planners and decision makers. They also claimed that much of the quantitative work was mere mathematical modelling based on poorly formulated problems, incorporating naïvely constructed objective functions, neglecting behavioural variables, and including politically and socially unfeasible alternatives. Such approaches often led to journal articles, but rarely to implementable solutions.

Public health journals certainly contain plenty of articles, it is true, extolling applications of systems thinking. But, in most cases, the best that can be said is that a *prima facie* case has been made for the type of application in question only with a specially concocted illustration using appropriate health data. Whether the work has in reality been an integral part of a genuine ongoing decision-making process may often be doubted. When the systems analysis is only hypothetical, all the other objections mentioned above can easily be made and are impossible to refute. But such criticisms, although valid so far as they go, really amount to saying only that genuine applications of systems analysis have not been made, and not that they are invalid in the sense of having been tried and found wanting.

Let us see therefore what many people currently regard as a systems approach to health problems. One source of information is the book *Systems Aspects of Health Planning* (edited by Bailey and Thompson, 1975). This contains the proceedings of a conference held by the International Institute of Applied Systems Analysis at Baden, Austria, in August 1974. The main areas covered by the conference were: (1) general concepts in systems approaches to health, (2) definition of health problems, (3) medical information systems, (4) resource allocation, and (5) planning and management in the health sector. While these headings are a broad guide to the subject matter, many of the papers, perhaps not surprisingly, did not fall squarely into a single classification.

Several contributions described serious attempts being made by governments, e.g. in Sweden and Czechoslovakia, to take a systematic and comprehensive approach to the urgent but highly complex problems of planning ahead for their countries' health services. In particular, two papers of considerable range and breadth were presented from the ministerial level itself by D. D. Venediktov, Deputy Minister of Health, USSR. These dealt with the general systems analysis of health services, and a systems review of the development of biomedical science and methods of coordination at both national and international levels. It is of course impossible here to go into details, and interested readers should consult the book referred to—bearing in mind that these papers are themselves no more than brief summaries of the actual work in progress at the national levels.

Generally speaking, the systems principles described in Section 8.1 have, in one form or another, been followed at least approximately. Most of the work is still in the early stages, and a lot remains to be done. Perhaps the most important aspect to be emphasized here is the following. In these papers deriving from real-life decision-intensive situations, there is clear evidence that some decision makers, at least, are fully aware of both the need for a systematic logical analysis of the administrative problems *and* the human needs for improved health services giving rise to those problems. The degree of success actually achieved in this work remains to be seen, but the potentialities appear to be considerable.

Some of the papers produced relatively sophisticated mathematical models, for example one by Fleissner describing an integrated model of the Austrian health care system. This modelling is essentially of the system dynamical type which we shall deal with more fully in the next section. The main point is that, given a number of structural and quantitative assumptions, some of which represent factors under control, forecasts can be made to predict what, approximately, is likely to happen in the future. In principle, this provides a basis for choice of strategies. In practice, however, things may be more complicated than that, and the author of the paper himself commented on the fact that: 'The absence of an overriding and driving goal within the model reflects the real situation in Austria. Goals vary across time and social class. The approach adopted was to distil a descriptive simulation of reality—in a sense, a status quo analysis. To have posited a hierarchical goal structure would have been to misdepict the system.' This clearly leaves some crucial loose ends to be tied up before the final value of this particular investigation can be assessed.

Other contributions with a strong mathematical basis (often implicit in the papers as presented) were those dealing with resource allocation problems, such as health manpower and mass health screening. This is what we should expect from more technically well-defined problems of an operational research type.

This is therefore considerable evidence that the total failure of systems analysis in regional health planning, allegedly observed by Shuman, Wolfe, and Speas, is beginning to be corrected. In many Ministries of Health nowadays one can find operational research or systems analysis groups (or their equivalents) which are in fact providing a direct service to highly placed decision makers, including the Ministers themselves. It is, however, significant that in some cases there appears to be no close link between such groups and those University departments one would expect to find providing strong technical support. Though this opens up another vitally important area of discussion, it would take us too far afield to analyse it here. We shall reserve further discussion to the next chapter.

Another example of a brief, but clear, statement of a systems analytic attitude of mind by a decision maker at the top of the health pyramid, is the Working Paper entitled *A New Perspective on the Health of Canadians.* This is issued by the Government of Canada and is signed by Marc Lalonde (1974), Minister of National Health and Welfare. It develops a conceptual framework of the health field, but makes no attempt to be dogmatic, or avoid controversy, 'because it is only through honest disagreement and warm debate that the broader issues of health can be clarified and further progress achieved.'

While the outlook is promising, it is only fair to say that really significant advances in the application of systems analysis still lie in the future. The solutions to many medical and health problems will be derived largely from more intensive work at the technical level. But what about some of the ubiquitous scourges described in Section 1.1? What about cancer, for example, where a relatively large proportion of the populations of most developed countries is affected sooner or later, and the same pattern is also beginning to arise in developing countries?

A vast amount of effort has been put into the fight against cancer on a world-wide basis over a great many years, but the results so far are a long way from what suffering humanity needs. True, some forms of cancer can be cured, and many advances have been made in alleviation and treatment. But far more needs to be done to prevent the occurrence of cancer in the first place. Considering the amount of research that has been done, and the numbers of people and financial resources involved, the successes, though real, seem relatively small. While it would be ridiculous to suggest that systems analysis could be the answer to this vast collection of interrelated problems, it does seem that a new, long, hard look at the total picture is required. And for this, systems analysis may be helpful.

Consider for a moment the range of problems occurring in the cancer field. These arise on many different levels, and in many different areas of study. For example, there are:

(*a*) the pursuit of epidemiological research, including geographical pathology;

(*b*) the prevention of carcinogenic hazards due to environmental, behavioural, and occupational factors;

(*c*) the screening of populations to detect cases at early pre-cancerous stages;

(*d*) the biological understanding of tumour growth and host response;

(*e*) improvement in techniques of early diagnosis;

(*f*) the development of chemotherapy, radiotherapy, and surgical techniques;

(*g*) the application of better combinations of the foregoing methods of treatment and prevention;

(*h*) the search for more cost-effective methods, and the use of cost–benefit analysis;

(*i*) the provision of better cancer health services;

(*j*) the establishment of better overall planning, coordination, and management for cancer control;

to mention some of the major aspects.

With such a diversification of component activities there is bound to be great difficulty in pursuing a well-balanced development. Moreover, it is easy to concentrate attention on specific, concrete activities like dealing with individual patients, investigating the properties of a long series of anti-cancer drugs, developing a new surgical technique, or studying mathematical models of alternative screening programs. Whereas to deal satisfactorily with the broader aspects of planning and overall policy is not only much more difficult: one also runs the risk of appearing merely to play with abstractions. Nevertheless, policy decisions *are* made, either positively or by default, huge sums of money *are* spent, and cancer *is not* yet under very effective control.

There is, I believe, a wide field of endeavour here which could be materially assisted by appropriate comprehensive, systematic forms of investigation of the total structure of existing activities. This needs to be pursued both nationally and internationally. Naturally, there are already many institutions in individual countries such as universities, hospitals, medical research centres, regional administrations, and ministries of health, all of which are contributing in some way or other. There are also several international organizations, including the World Health Organization, whose specific role is to investigate the balance between the innumerable activities in progress and to devise if possible appropriate means of redressing any glaring gaps that may exist in the total world picture, that is, to improve overall coordination and organization.

What is needed is not, of course, any attempted comprehensive systematization of *all* known data about cancer (itself capable of classification into at least a hundred distinct types of tumour growth requiring individual treatment), but a judicious sketching out of the main areas of world activity, together with the chief interactions between such areas. In other words, we need a situational analysis in broad terms, a suitable conceptual framework etc., along the lines already discussed in Section 8.1. Fortunately, a good deal of relevant material already exists, since many areas have already been surveyed and reviewed in the literature by those most competent to do so.

Of course, these remarks are intended only as suggestions for possibly fruitful investigations into an area of major consequence to mankind in which systems analysis might be able to make a significant contribution. It is evident from the Conference on Cancer Information Systems that took place in Moscow at the beginning of December 1975, jointly sponsored by the International Institute for Applied Systems Analysis and the World Health Organization, that many groups of people, at both national and international levels, are beginning to think on these lines and to explore the potentialities. The next few years will show whether something worthwhile can be developed, though sustained long-term efforts will be required for genuine achievements.

8.3 SYSTEM DYNAMICS

The term 'system dynamics' has a number of connotations. First of all, it can mean the dynamics of any system of interrelated components having a sufficiently well-defined set of operating mechanisms. There is in fact a whole spectrum of possibilities here. We could start with the special case of a system having only one component—hardly a system at all. Take, for example, a homogeneous population of individuals in which the population size is large enough for it to be considered as a continuous variable. This population can be seen as contained within a single box. New recruits are added by birth at a continuous rate proportional to the population size, and losses occur in a similar way. There is thus a net rate of increase which is positive or negative depending on whether births exceed deaths or *vice versa*. A single deterministic differential equation, as in equation (3.1) of Section 3.4, then represents in a nutshell the whole process.

Again, in communicable disease theory, the simplest possible epidemic model is one involving no recovery or loss of infectivity. We can then consider a single box containing infectious individuals which is continually augmented by new infection. When removal from circulation is allowed there are two essential boxes: one for the pool of available susceptibles and one for circulating infectious persons. Transfers occur from the first to the second box, while the latter incurs removals as well.

Additional boxes can be introduced into the epidemic model to allow for different immunity states, or different transfer rates for different age and sex classes. And, of course, the one-box population model can also easily be modified to allow for age and sex structure. In fact, almost any special distinguishable features of such processes can be incorporated in the model by the use of appropriately chosen boxes with suitably defined transfer-rates. It should not be forgotten, however, as discussed earlier (see Section 5.4), that we may bring in the extra complexity to achieve greater realism only to find that we cannot determine the extra parameters involved and cannot effectively test the model against actual data. As usual, the degree of complexity introduced should be no more than is strictly necessary.

However, in many areas the processes studied may exhibit a considerable amount of well-documented complex structure that simply cannot be ignored. A relatively complex model with a large number of boxes, or compartments, is then inevitable. Many physiological systems are of this type, for example. When the basic treatment involves continuous flows or transfers, as above, the model is *deterministic* in character. The essential dynamic mechanisms of the whole system can then be expressed most compactly in terms of a set of differential equations. The solution, or solutions, of these equations, when appropriately reinterpreted in practical terms, describe in summary form what can be said about the overall behaviour of the whole system. All kinds of deductions can be made, providing general insights as well as practical implications. New hypotheses can thus be generated, and tested against observational data. Fresh evidence for or against the model is obtained accordingly.

When the processes being modelled contain major probabilistic elements that cannot be averaged out, as with chance infections in small epidemics, these must be incorporated in the model. We then have a more sophisticated *stochastic* model (see Section 4.5). Again, the essential dynamics can be expressed in terms of a set of differential equations, though the latter will now be much more complicated as they have to deal with the changing relationships between large numbers of individual probabilities. Sometimes specialized techniques, like the use of generating functions, will allow the mass of differential equations to be replaced by a much smaller set, and, if we are lucky, perhaps by only a single equation.

So far, these general ideas are very much the same as already discussed in earlier chapters (especially Chapters 3, 4 and 5). However, once the facility of handling the larger systems is achieved it is natural to extend the areas of application. This is particularly true when purely mathematical methods of analysis and solution are supplemented by computerized investigations (see Sections 6.5 and 7.3). While we are still dealing with the natural systems in the areas of traditional scientific endeavour, e.g. physics, physiology, biological populations, etc., we are on fairly firm ground. Theory and mathematical speculation can sooner or later be checked against observation or experiment. But as soon as we pass to the more complicated and changing problems of society, with its manmade systems, such as arise in commerce, industry, and social institutions, new difficulties arise.

The 'natural' world is to a great extent given, and we are constrained to accept its laws and modes of functioning. Any intervention or control that we achieve has to be effected within these contraints. But social, manmade, processes are in principle more plastic. Their structure is harder to pin down and is more liable to apparently arbitrary change. It is against this kind of background that we must consider the validity of the system dynamics of Jay W. Forrester, whose major books, *Industrial Dynamics* (1961), *Urban Dynamics* (1969), and *World Dynamics* (1971), have been the centre of so much attention and discussion in recent years. The general relevance to health is immediate, because health service systems themselves are of precisely this type, as are many of the other systems occurring in sectors with which the health sector interacts, such as education, agriculture, water resources, energy production, economics, etc. (see Section 1.4). So far as logical form is concerned, this kind of system dynamics is hardly to be distinguished from the modelling of dynamic processes occurring in biology or epidemiology, but as indicated above there is a shift in the type of subject matter dealt with. So for reasons partly of history and partly of convenience, system dynamics has appeared as a new discipline in its own right.

Forrester's *Industrial Dynamics* is a classic masterpiece of lucidity and scholarship, combining theoretical insights with practical relevance. True, Forrester rarely acknowledges his indebtedness to other writers, but his energy and enthusiasm have certainly served to establish a method of investigating large-scale, complex processes that has now become one of the essential techniques. In Forrester's own words, 'Industrial dynamics is a way of studying the behavior of industrial systems to show how policies, decisions, structure and delays are interrelated to influence growth and stability. It integrates the separate functional areas of management—marketing, investment, research, personnel, production and accounting. Each of these functions is reduced to a common basis by recognizing that any economic or corporate activity consists of flows of money, orders, materials, personnel and capital equipment.'

All this is systematized by building appropriate models, and is developed using special techniques derived from information–feedback theory dealing with the goal-seeking, selfcorrecting mechanisms operating in business systems, the theory of decision making (especially in military contexts), and the opportunities offered by modern large-scale electronic computing facilities. The possibility of integrating all these new developments meant that it became feasible to handle effectively the greater complications occurring in the dynamic systems of industrial and economic activity.

Forrester later pointed out that 'industrial dynamics' was too limited a term, since the methods expounded would apply to any complex dynamic system, regardless of its field of application. Hence the introduction of the term 'system dynamics'.

A whole chapter of *Industrial Dynamics* is devoted to the problems of judging model validity and how to carry out appropriate practical tests. It is pointed out that *quantitative* validation must be done where possible, and, when justified, in cost–benefit terms. Nevertheless, because so much of the structure of large complex systems is qualitative, we cannot expect in general to be able to carry out the prediction of detailed events. Thus it is categorically stated that, 'System models should predict and reproduce the behavior or character of a system, not specific events or particular, unique sections of actual system time history.'

Although there is a good deal to be said for this view, it appears to go too far in avoiding rigorous submission to the normal scientific standards of verification. Admittedly, if a systems model is used to influence decision making then we only know what happens in relation to the policy chosen; there are no historical records relating to the alternative policies that were not chosen. Nevertheless, we can certainly see whether the detailed predictions of the model are verified or not in the special circumstances resulting from a chosen policy. Moreover, there are a great many ways in which standard statistical tests can be employed to check the detailed implications of any given systems model against observational data. And if there is insufficient evidence for an underlying model, why should we believe in it?

This question of validation is in fact crucial to the whole technique of using systems models and system dynamical approaches, and has given rise to much heated controversy in regard to the later developments mentioned below. In his subsequent major publications, Forrester tends to give the subject little prominence. Thus, his *Urban Dynamics* extends the earlier ideas to problems of urban behaviour and urban policies, by constructing systems models of supposedly fundamental urban processes.

It would take us too far away from our main theme to examine this work in detail. But it is important to point out that Forrester not only draws attention to the fact of the 'counter-intuitive' or 'counter-productive' effects of many economic and social policies (i.e. policies whose actual effects are the opposite of those desired), but also emphasizes that such results are confirmed by the system dynamics of the basic urban processes. From this analysis 'it emerges that the fundamental cause of depressed areas in the cities comes from *excess* of housing in the low-income category rather than from the commonly presumed housing shortage.' He suggests a reversal of present practices so as simultaneously to reduce the aging housing in cities and to allocate land to income-earning opportunities.

Although closely argued, these are large claims with far-reaching social, administrative, and political implications. The point we are making is that in such sensitive areas we have to be extremely careful about jumping to conclusions based on what are, to most people, incomprehensible computerized calculations applied to a complex model whose structure might well be disputed if only enough people could understand what it was. While there is no reason whatever to doubt Forrester's genuine enthusiasm and sincerity, the ease with which the disaffected can disavow the whole approach has done considerable harm to its general acceptance. The importance of careful validation at all stages cannot be overestimated.

The difficulties are further aggravated in *World Dynamics*. This literally deals with problems on a world scale by combining submodels for population, natural resources, capital investment, capital devoted to agriculture, and pollution, into one grand super-model. There seems little doubt that Forrester produced a *prima facie* case for a re-examination of a number of major issues of vital concern to mankind. He rightly pointed out that certain computer print-outs 'should make us cautious about rushing into programs on the basis of short-term humanitarian impulses. The eventual result can be anti-humanitarian. What is the proper trade-off between benefits in the near future in exchange for distress in the distant future?' However, one further reads that, 'Rising pressures are necessary to hasten the day when population is stabilised. Pressures can be increased by reducing food production, reducing health services, and reducing industrialization. Such reductions seem to have only slight effect on quality of life in the long run.'

If the assumptions of the modelling are all more or less correct, the structuring of the system roughly what we would all agree to, and the mathematics and computerization all without fault, then it *might* be that Forrester's conclusions are simply hard truths we have to accept in order to survive. Nevertheless, the validity of the system is hardly mentioned, even though answers to existential questions of extinction or survival are based on it. Moreover, from the point of view of our own subject, medicine and public health are explicitly mentioned as being included in the world dynamics system as a part of the industrialization process. So some care in interpretation is needed.

This work was continued and further developed under the auspices of the Club of Rome, and culminated in the now celebrated, if not notorious, *The Limits to Growth* by Meadows, Meadows, Randers, and Behrens (1972). The storm of objections raised to this widely read work, with its clearly expressed predictions of probable doom to mankind in the near future, are now well-known. On the whole, on might have supposed that a number of serious issues had been brought out into the open, as in *World Dynamics*, and that the apparent consequences of increasing population, industrialization, and pollution, together with dwindling resources, needed urgent re-evaluation. The enraged protests in otherwise respectable scientific journals invited the conclusion that many scientists could not accept the notion that modern science was not all-powerful and might easily fail to solve present world problems simply by making fresh discoveries in the future. In fact there was a strong presumption that new ways of living and organizing society might become essential if disaster were to be averted.

Of course, a reasonable scientific attitude would be to examine the assumptions underlying *The Limits to Growth*, devise methods of checking the validity of the structure of the modelling, modify the hypotheses where necessary, etc., until some generally acceptable conclusions had been reached. Presumably the emotional outbursts that were generated occurred because already existing deep-seated anxieties were sparked off. My own personal belief is that *The Limits to Growth*, for all its obvious oversimplifications and shortcomings, has provided an inestimable service to mankind in ventilating major human issues in terms of a tentative, quantitative conceptual framework. Nevertheless, a lot of further work is required to provide a generally acceptable rational basis for both national and international policies.

A second report to the Club of Rome, *Mankind at the Turning Point*, by Mesarovic and Pestel (1974), has in fact produced a less dramatic and more optimistic view. The new world model is much more complex and sophisticated, and it appears that some '100,000 relationships are stored in the computer.' Moreover, in addition to the basic world system computer model, there is also a decision-making model. Instead of the older, simpler, more mechanistic, type of modelling, as exemplified by Meadows *et al.*, Mesarovic and Pestel envisage an interaction between the systems analyst and the computer. This constitutes a 'heuristic' or 'interactive' approach to decision making, in which the analyst becomes more actively involved in the evolution of the system itself and contributes to the restructuring of the system in an attempt to produce a more stable overall equilibrium (see also Section 5.4).

These vital issues must continue to be debated until satisfactory guides to policy making emerge. But the world is a big place, and it is hard to resolve questions as to the validity of the whole approach in such a large-scale setting. In the more restricted area of the health sector itself, but not forgetting its interactions with other sectors, there is much to be said for further exploration of the system dynamics approach. If the legitimate qualitative objective of improved health for the world, as discussed in Chapter 1, is to be realized, then a great deal has to be done in assessing the feasibility of the various subgoals mentioned in Section 1.4, particularly with regard to the constraints imposed by the hard facts of health economics. There is a vast interrelated system of interacting component activities, combining straightforward quantifiable processes along with human values and aspirations.

The right kind of system dynamics could probably make material contributions to these problems. But it would probably have to be developed first at national levels, where both consumers and decision makers could cooperate in developing generally acceptable models of the intricate processes involved, and where the publicly recognized validity of the approach could be guaranteed by appropriately designed methods of testing the models against observable facts.

We have already referred in the previous section to the conference proceedings published in *Systems Aspects of Health Planning* (edited by Bailey and Thompson, 1975), and to the paper by Fleissner on an integrated model of the Austrian health care system. The latter is a good example of a tentative but serious system dynamical approach, but it is evident that a lot more work is required before an instrument can be developed that will be capable of close integration with the policy-making process. At the same time, careful examination of the details of the system shows that there are many components of a technical kind that readily lend themselves to individual validation.

Generally speaking, we ought to be able to build up more acceptable system dynamics models in the health field by ensuring piecemeal validations of all technical components plus a number of critical tests of specific predictions. When the underlying models can be demonstrated to have passed such tests in a sufficiently convincing manner, they will be usable with a reasonable degree of confidence for predicting total system behaviour. This is what planners need in order to choose feasible options on the basis of expected health, economic, and environmental consequences.

In any case, there are already substantial benefits at an earlier stage before a really operational system is evolved. The mere act of explicitly stating objectives and assumptions, and clearly defining how predictions are to be made and decisions reached, all amounts to a valuable process of clarification. An attempt to improve on this standard systems analytic approach by further quantitative dynamic modelling can then sharpen understanding still further. If mistakes occur, they can be recognized. Lessons can be learned and new knowledge acquired and transmitted to others. Planning skills can thus be more widely disseminated on a rational basis, instead of remaining the intuitive private property of self-styled experts.

CHAPTER 9

Teams, Projects, and Organizations

9.1 INTRODUCTION

'Die Philosophen haben die Welt nur verschieden interpretiert,' runs Karl Marx's eleventh Thesis on Feuerbach, 'es kommt aber darauf an, sie zu verändern.' We have already stressed, especially in Chapter 2, the necessity for validating theoretical speculations by reference to practical observation or experiment. Marx's dictum goes further: it implies that such speculations are ultimately justified only in so far as their consequences can be dynamically incorporated in the changing and developing life of the community. This raises a number of fundamental issues connected with the possibility and desirability of directing and planning scientific research.

It is often pointed out that the tremendous outburst of intellectual activity amongst the Ancient Greeks occurred quite spontaneously without any conscious planning or organization, and that this flowering of human thought has continued to be a source of both inspiration and practical value ever since. This much is undisputed fact. But it is frequently argued further that all creative thought and research should continue to be completely untrammelled by any attempt to direct it into specific channels. However, realistic conclusions cannot be reached without careful reference to the historical context.

In the first place, no one before the Greeks had ever really indulged in such unfettered speculation over the whole range of human interest and activity. Any philosophical enquiries at that time, if pursued with determination and enthusiasm, would have been almost certain to produce much of lasting value. And, of course, it must be remembered that while certain Greek theories were eventually proved to be more or less correct, such as the primitive ideas of Democritus on the atomic structure of matter, many others turned out to be false. The evidence available at the time was often inadequate to determine which theories were right and which were wrong. What happens is that we award recognition retrospectively to those thinkers who happened by chance to be correct. In fact, credit is really due to all those whose thought showed inspiration, ingenuity, and imagination, regardless of its truth. However, in a modern advanced community, capable of conscious and critical attention to its intellectual activities, the situation is quite different. It would simply be very wasteful, and possibly dangerous, not to have some degree of planning and organization, though the precise form that these should have is a matter of considerable controversy. Free speculation is highly desirable; so is utilitarian application. The relation between the two is of critical importance, and is in one way or another the primary aspect of most of the ensuing discussions in this chapter.

In the second place, most scientific research is very expensive. Even the cost of supporting theoretical mathematicians is not inappreciable, where the main budgetary items may be no more than salaries, office accommodation, library facilities and some clerical assistance. As soon as laboratories and special equipment are involved there may be no limit, other than that set by economic necessity, to what may legitimately be asked for. Nuclear physics and space research are obvious cases in point. Even biomathematics, which we can regard as covering the whole field of mathematical applications to biology, medicine, and public health, can lead to quite large expenditures as soon as, for example, extensive statistical surveys are undertaken, or practical operational trials are set up, especially when these have to be supported by large-scale computing facilities. Of course, relatively speaking, the whole field of biomathematics is at present considerably underfinanced, but to carry through much of what can already be foreseen quite large sums of money would be required. The general result is that a very high proportion of scientific research is now supported, directly or indirectly, by government agencies or industrial concerns where utilitarian motives are paramount. As a rule, pure research is promoted as well as applied research and operational research, but naturally any proposed pure research projects are more likely to receive support if they appear to have some bearing on specific applications.

It is obvious that decisions have to be made at a high administrative level to spend the available money on one project and not on another, and considerable restriction of the freedom to pursue independent scientific enquiries inevitably follows. Even supposedly autonomous research institutions are to some extent bound by their constitutions or basic terms of reference, and are further restricted by the governments and foundations to whom they apply for grants. Again, there is usually much greater freedom to follow independent research investigations in a university setting, though it may still be necessary to satisfy the officials of some grant-giving body that one's work is worth supporting.

The ideal of complete freedom in research is accordingly unattainable. Every research workers is in fact limited by the job he can find, the people he has to work with, the equipment he can obtain, the sources of economic support that are available, and indeed the whole intellectual and social environment within which he finds himself. Within these limits, however, a certain degree of partial freedom is possible. It is clearly of the utmost importance to preserve and foster this freedom if the spirit of adventure and free enquiry are not to be stifled by unavoidable constraints. To the writer it seems unlikely that this will be achieved by a *laissez-faire* attitude of leaving matters to themselves. For then some form of planning and organization, possibly of an undesirable kind, will continue to be imposed by various authorities because of existing circumstances. The only realistic procedure would seem to be to work consciously and deliberately for forms of organization that will guarantee the maximum amount of freedom within the given environmental framework.

The above remarks apply of course to scientific research in general. We in this book are primarily concerned with the implications for the broad field of biomathematics, especially in regard to the medical and health applications already discussed in previous chapters. A more detailed assessment of the whole situation will be undertaken in Section 9.4, and certain proposals will be made for an integrated approach to the practical implementation of biomathematical science. In the meantime we should take a closer look at the range of activities involved.

Thus, in a typical university department concerned with biomathematics, biometry, biostatistics (or whatever term one chooses to use for the collection of biological, medical, and public health mathematical activities under discussion), the work may quite conveniently be classified under the three headings of teaching, consultation, and research. These mutually dependent aspects are discussed more fully in the next section. Some biomathematical research can be pursued in relative isolation, but as repeatedly emphasized in this book it is in general necessary for close contact to be maintained with the real world of experiment and observation. Biomathematicians should therefore be intimately associated with the work of appropriate biological and medical institutions, since these frequently need mathematical and statistical advice. But if the integrated approach we have been recommending is to be developed, then far more carefully planned cooperative research is required. This topic is elaborated upon in Section 9.3. Again, when applications involve organizational processes the key activity may be principally of an operational research nature. This involves cooperative investigations of a more difficult type, since a greater number of experts with more diverse backgrounds will usually have to take part. Section 9.4 is devoted to a specific examination of the leading problems in this area.

It is quite possible, and sometimes even desirable, for whole institutions to concentrate mainly on a single type of activity, such as teaching, consultation, pure research, applied research, or operational research (or any other subdivisions that seem appropriate). But since all these aspects are closely related, and advances in one area may radically influence the work in another, it is worth giving serious attention to the problem of ensuring adequate mutual stimulation and cross-fertilization between the different divisions. In a small institution emphasis may have to be made on only one or two aspects, the rest being excluded or playing only minor roles, though with larger organizations some dynamic balance may be possible over the whole spectrum of activities. This should enable new ideas to be taken up, developed, and communicated to those who are concerned with teaching and consultative advice more quickly than would otherwise be the case.

In Sections 9.2–9.4, we shall deal with the general problems of biomathematics as a whole, reserving Section 9.5 for discussion of the more specific needs of applications in the health field.

9.2 TEACHING, CONSULTATION, AND RESEARCH

As already emphasized in the previous section, if the mathematical approach to biology, medicine, and health is to have any fundamental importance it must be seen to possess not only scientific validity but also valuable implications for the community in general. The writer believes that both these criteria are fulfilled. Consequently the fruits of biomathematical research must be made available to those who can apply them to their own work, as well as to those whose job it is to communicate knowledge to others. The first aspect we have broadly referred to under the heading of 'consultation'. This may entail supplying mathematical, statistical, or computational advice and cooperation to other research workers to the extent that joint research is involved, or it may merely mean that cookbook techniques are made available to be used routinely like any other piece of scientific equipment. The second aspect may be categorized as 'teaching'. Let us consider this matter first.

Teaching

We are concerned here with the communication of the mathematical approach to biology, medicine, and health to those who wish to learn more about it, as distinct from the teaching of specific subdivisions such as certain kinds of applied mathematics or computer science which may be acquired at school or as university 'subjects' at many different levels. The problems involved are fairly complex. But, broadly speaking, if biologists, clinicians, and health administrators are to increase their understanding and practical grasp of mathematical techniques then the latter must be taught with sympathy and enthusiasm with specific reference to the contexts concerned.

The difficulties in the more exact sciences are not quite so acute, since a great deal of applied mathematics, at least as taught in Britain, automatically entails even at school level a considerable range of mathematical physics such as the motion of planets round the sun, elementary ballistics, heat conduction, electromagnetism, etc. Those who subsequently specialize in theoretical physics or mathematical physics may therefore have been exposed to the basic ideas and principles for quite a long time. And even the more experimental physicist, especially if he has done some applied mathematics at school, will be quite familiar with the general implications of the mathematical approach.

On the other hand, those who take up biological, medical, and public health subjects may do so in part because they are less interested in, or less able at, mathematics. This of course intensifies difficulties in much advanced work since, as we have already stressed, biological material is frequently so complex and variable that elaborate mathematical models are required for satisfactory representation, and their handling and elucidation may necessitate highly sophisticated mathematical and computational procedures. It is desirable, therefore, that those who do biology and premedical subjects at school, with a view to specializing later, should also receive an adequate amount of biomathematical training. This training should be an integral part of the courses followed, and not just an independent subject whose essential relevance could easily be ignored. How much training would be 'adequate' cannot be dogmatically specified. Some, whose intrinsic mathematical competence is not outstanding, might achieve little more than elementary algebra but should at least understand what mathematical models are, how probability ideas provide valuable clues to grasping biological phenomena, and how massive quantities of data and complicated calculations can be handled by computers. Others might be expected to go further and, for example, learn to handle elementary differential equations and to perform some of the simpler procedures in numerical analysis and computer programming.

The trouble is that many students who are in fact capable of acquiring some basic skill in mathematics, statistics, and computing are not encouraged or given the opportunity to do so until they are already at the postgraduate research stage. They may need a fair amount of biomathematical knowledge in a hurry in order to handle the subject of their theses, and the attempt to acquire quickly a possibly unfamiliar method of thinking and the specialized techniques that go with it may easily present an intolerable burden. All this could be avoided if a graduated program of biomathematical training were instituted, beginning with the basic elements at school and continuing with more advanced work throughout the undergraduate period. Courses should be specially tailored to meet the needs of those who will be primarily biologists and those who will follow the medical curriculum. There is a good deal of overlap, but several specialized requirements are different. For instance, doctors need to know more about operational research as applied to hospital functioning and the use of statistical concepts to handle ethical problems in therapeutic trials, while biologists might prefer more emphasis on classical applied mathematics and the computing of solutions to elaborate mathematical models (with or without probability elements).

In the long run such a process of graduate training over an appreciable period of time would enable the philosophy and techniques of biomathematics to be universally acquired. In the short run, however, a variety of *ad hoc* courses are needed. Thus lectures on statistical methods are frequently given both to undergraduates and research students, in the hope that they will rapidly acquire at least the ability to analyse their experimental data more efficiently. While this is a valuable and necessary expedient, it is usually very difficult to inculcate in a short time the proper attitude of mind without which the correct interpretation of results may easily be missed. Again, both medical students and practising clinicians can be introduced to the notions and implications of operational research and data processing. But unless the whole quantitative approach is already a familiar one, great difficulty may be found in assimilating the new ideas, and they may fail to be used on the very occasions when they are most needed. In times of stress and crisis new modes of thought tend to be forgotten, and only long-accepted attitudes made use of. Hence the importance of inculcating the biomathematical approach slowly and progressively over an extended period of time.

So far we have been discussing the teaching of biomathematics to those who were or would be primarily biologists or doctors, and this is perhaps the greatest need. At the same time much biomathematical work is currently done by people whose main expertise is in the mathematical field but who have developed special interests in biological and medical subjects. Whether this is entirely satisfactory is a matter of great debate and we shall refer to the problems entailed again later. Although it is possible to envisage biologists, doctors, and health administrators becoming so competent at biomathematics that they could do all the required work themselves, thus combining several specialized skills within single individuals, it seems much more likely to the writer that it will always be more convenient to have biomathematical experts to concentrate on and develop the subjects involved.

It follows therefore that we must also cater for the needs of people whose work is going to be primarily biomathematical. In general they are likely to have had a strongly mathematical background, as specialized knowledge is usually required for most advanced work and research. If they also have qualifications in biology and medicine, so much the better, but for some time to come individuals with multiple accomplishments are likely to be a small minority.

The commonest source of biomathematicians at present is via a training in mathematical statistics. While it is encouraging that statistics is now starting to be introduced as a school subject, it is frequently not taken up until the postgraduate stage when newly qualified mathematicians read for a postgraduate diploma or research degree in statistics. This tends to give a considerable bias towards pure mathematical statistics. The balance is redressed to some extent by the inclusion of special courses (and examination papers) on biomathematical subjects, but these may easily be regarded as mere illustrations of the application of mathematical theory. Only a few of these students seem ultimately disposed to embark on careers in biomathematics as such. To a great extent this is probably due to their being introduced to the subject rather later in life. When it comes to choosing a job it is easier and safer to take a post where the main requirement is merely for general statistical advice and routine computational assistance.

The situation is aggravated by the tendency of many professional mathematicians to regard research in pure mathematics as having the highest status, while only the second-rate go into mathematical statistics. Applied statistics is on a lower level still, followed by operational research. This intellectual snobbery has the unfortunate effect of diverting many first-rate mathematicians from subjects in which their skills are badly needed and in which they would find a challenging and interesting variety of problems.

Some universities have first degrees in statistics, and a good deal of pure and applied mathematics may be included in the relevant courses. As time goes by much work formerly done only at graduate levels is being introduced to undergraduate students. This tendency is all to the good. Further, as with the teaching of biomathematics to biologists, many mathematical statistical notions can also be introduced at school level.

With more probability and statistics being taught in schools, it should be easier to incorporate many biomathematical topics in the applied mathematics work. Instead of concentrating largely on physical applications, those with the requisite abilities could be learning some elementary mathematical genetics or studying the simpler aspects of stochastic population behaviour.

What we are specifically recommending is that biomathematical subjects should be taught and encouraged at all levels from school onwards so as to provide a gradual but progressive education. Training must be adapted to meet the requirements of different kinds of student, depending on whether their primary abilities are in mathematical or biological fields. Unless teaching is commenced at a fairly early age the special ideas and concepts involved in biomathematics will not become an integral part of the subsequent intellectual equipment of the students, and only a minority will be able to handle the salient techniques with appreciable facility.

Consultation

As biological and medical work becomes more complex, and as the ideas discussed in the earlier chapters of this book become more widely accepted, so the need for expert advice and assistance will grow. We have already suggested that for some time to come biologists and clinicians will prefer to have professional biomathematicians undertaking this kind of work rather than attempting all of it themselves. At the same time it is to be hoped that both groups will come closer together, so that for example biologists learn more biomathematics and biomathematicians learn more biology. A certain division of labour, here as elsewhere, would seem to be convenient and economical.

A typical and familiar situation is that a research biologist in the field or laboratory is undertaking a fairly complex experiment involving many factors, and he feels the need of some advice on the general pattern of the experiment as well as with the calculation of results. We have already looked at the reasons why 'experimental design' is so important in Section 4.4, and it is now widely appreciated that if the correct statistical procedures are not followed much arduous and expensive experimentation may be wasted. Questions arise therefore as to what kind of advice the biologist should be given, what knowledge the statistician should have, what degree of cooperation is desirable, and so on.

If the biologist is familiar with the principles of experimental design he may require no more than the routine computation of results from a design that he has already chosen and put into operation himself. In this case the statistician is performing only a minor service function. But quite often the choice of design is not an easy matter, and depends on considerable mathematical knowledge plus an understanding of the biological hypotheses to be investigated. The statistician's role is then of considerably greater importance and amounts at least to essential technical advice and assistance. And on occasion the statistician may play an even more intimate part in carrying out the research. It may happen, for example, that the various hypotheses which the biologist wishes to test cannot be examined within a single experiment, or that a new kind of experimental design must be developed, or that the hypotheses must be modified and somewhat different questions asked, or different mathematical models adopted, etc. When this happens a high degree of cooperation between biologist and statistician is necessary, each understanding in considerable measure what the other is trying to do. The research is then really a joint enterprise (the subject of cooperative research in general will be discussed more fully in the following section), each expert making a distinctive contribution to the work. When it comes to the stage of publishing results in a scientific journal the paper should be presented as a work of joint authorship, whereas in the two previous cases no more than a suitably worded acknowledgement would usually be required.

Questions of scientific publication are important to biological statisticians and biomathematicians, since career advancement depends to a great extent, as with other scientists, on actual authorship of research papers (and not on mere honorable mentions). All too frequently their contributions to research are overlooked and they do not receive the recognition due to them.

The above remarks in the specific context of statistical experimental design apply with equal force to any sort of biomathematical work. It is evident that if biomathematicians are to be capable of a close degree of cooperation with biologists and clinicians then their training must have been specially designed to fit them for this task. Some aspects of training in biomathematics have already been mentioned in the previous subsection on teaching. A further matter of importance is the extent to which consultant biomathematicians should teach, and conversely. There is also the position of primarily biomathematical research to be considered. This we shall examine more fully in the next subsection.

Broadly speaking, it is desirable that everyone should participate to some extent in all of these three activities, the precise proportions of time spent on each depending on scientific need and personal inclination.

Although some consulting work is of a straightforward service kind, it is surprising how often discussions between biologists or clinicians and biomathematicians reveal unsuspected difficulties requiring special skill for their elucidation. The process of cooperative research is then likely to involve an appreciable degree of mutual education. From the biomathematician's point of view, not only is he acquiring valuable biological or medical knowledge, but he is in fact conducting informal teaching sessions in his own subject. If there are several people requiring constant advice it may well be worthwhile instituting more formal courses of lectures on appropriate topics. These should them facilitate the consulting work, while the latter suggests the form the lectures should take. Again, teaching by itself is apt to be dry and lifeless if not related to reality. It is therefore highly desirable that all teachers should undertake some kind of advisory work, whether they operate at school or postgraduate levels. This mutual dependence of consulting and teaching indicates that the two activities ought always to be considered together, the exact relationship being contingent on circumstances.

Research

While teaching and consultation basically entail the communication of special knowledge, research is the acquisition of this knowledge. Some people carry out research because they like it, others have it thrust upon them. Biomathematical teaching and consultation as such are basically crafts, and require a high degree of comptence and long experience to be really successful. At the same time students' questions or scientists' problems may easily occasion special enquiries of a research nature. Teachers and consultants therefore find themselves drawn into research even if this is not their main interest. Conversely, those whose primary activity is the carrying out of research will usually want to communicate their findings to others and to train the next generation of research workers. Consulting and teaching thus arise as almost automatic concomitants to research. Again, teaching is enlivened and enriched by close contact with research, and is seen to have practical value through the real-life applications occurring in consultant work.

As with other sciences, research in biomathematics is what provides the subject with excitement and the spirit of adventure, whether it is concerned with furnishing greater insight into existing knowledge or whether it is occupied with exploring hitherto uncharted areas. Biomathematical research may be pursued by relatively isolated individuals or by teams of investigators; by those who are working nearly full time on it or by those who are chiefly occupied in teaching or consultation; by members of universities, government agencies, or independent institutes; and so on. The range of possibilities is considerable, though the actual number of openings is at present still regrettably small.

The previous chapters give some idea of the general range of biomathematics, including a number of special topics presented in depth by way of illustration. Free enquiry is the essence of research, although it is often subject to certain overall restrictions regarding the general field of investigation as discussed in Section 9.1, as well as to the formal logical demands of the scientific method analysed in Chapter 2.

Research in biomathematics may be conveniently classified under four main headings provided by some of the titles of previous chapters. This is of course only one method of categorization, but it is specially germane to the present chapter where we are principally concerned with problems of organization. (We could also classify by subject matter, techniques, institutions, etc.) First, there is the basically *mathematical* aspect since we want to be able to apply, manipulate, and develop any of the concepts and techniques found in pure and applied mathematics. Secondly, *probability and statistics* occupy a special place because of the universal variability of biological phenomena. Thirdly, *operational research* (including *systems analysis*) is an essential category since the application of biomathematical knowledge frequently calls for the optimizing of organizational systems if reasonable efficiency is to be maintained. Fourthly, *computing* is a specialism which none of the previous three departments can nowadays afford to be without, because of the massive quantities of data to be handled and the intricate computations to be performed.

Any single investigation may cover all four of those aspects, and each aspect may necessitate a high degree of individual expert knowledge or special technical facilities. It is common therefore for many researches to involve several people. One of these may be the real investigator responsible for the project, while the others may only provide expert advice over limited areas. Alternatively, the project may be a genuinely cooperative enterprise with several research workers making key contributions. (The latter situation is so common in biomathematics that an amplified discussion is undertaken in the next section.) Thus in creating an environment within which research will flourish, due attention must be given to ensuring that the four aspects of biomathematics cited above are adequately represented. Otherwise an investigation may fail or be misdirected because a vital element in the process is not available at a sufficiently advanced level of competence.

Of course we must cater, not only for the prosecution of research itself, but also for the encouragement and training of new research workers. The general dissemination of biomathematical concepts and skills is primarily a teaching activity, as discussed above. But the initiation into research requires close association with, and participation in, actual ongoing research. Adequate provision must therefore be made for research students to have intimate contact with a wide range of biomathematical research investigations, to cooperate in some of those themselves, and to develop their own individual lines of enquiry. The organizational aspects of this matter will be taken up again in Section 9.4.

9.3 COOPERATIVE RESEARCH

Although biomathematics makes abundant use of results and techniques taken from pure mathematics, it is itself an essentially applied subject. As we have already remarked in Section 9.2 and elsewhere, much classical applied mathematics is taught in such a way as to incorporate a wide range of physical contexts. Applied mathematicians can therefore frequently carry out important research in these fields with a relatively small degree of immediate cooperation with other scientists. Also, it often happens that the physicists themselves are sufficiently able mathematicians to execute relevant mathematical research with little or no outside assistance. In the biological and medical sciences the situation is quite different. Most mathematicians acquire no appreciable biological or medical knowledge as an integral part of their main training, and the majority of biologists and clinicians have no specialized mathematical skills. Cooperation between experts with widely differing backgrounds is accordingly an essential feature of nearly all biomathematical research.

There are situations where the cooperation required is fairly minimal. Thus a biologist with some mathematical ability may develop quite accurately the partial differential equations describing a complicated physiological process, but may be unable to derive a solution. The problem could be handed directly to a willing mathematician with a simple request for 'the answer'. Such a procedure may turn out to be satisfactory if no difficulties arise. The mathematicians's role has then been of a primarily service nature, and no real scientific cooperation is needed.

However, it may easily happen that the equations require certain additional conditions or assumptions, or are difficult to solve in precisely the form presented. The mathematician may then introduce further limitations or modifications so that he can proceed. These variations may or may not be in the spirit of the original biological problem. If they are not, a great deal of intricate but useless mathematical investigation may be undertaken deriving the correct solution to the wrong problem. How is the mathematician to tell what is ultimately biologically permissible? Only by being interested in the biological problem itself, and becoming acquainted with all the relevant details. Close cooperation between mathematician and biologist is desirable from as early a stage in the research as possible and should continue through to its completion. In this way the thinking of each specialist can inform the activities of the other. The biologist should be prepared to adapt or modify his concepts and hypotheses in accordance with what he sees is mathematically or computationally feasible, while the mathematician will be prevented from going off in the wrong direction.

A similar state of affairs was mentioned in passing in the part of Section 9.2 dealing with 'consultation', where the case of specifically statistical advice was used as an illustration. Examples of many different kinds could easily be drawn from the whole field of biomathematical subjects. We may well ask therefore whether any special steps should be taken to ensure a proper degree of cooperation between mathematicians and scientists.

To a great extent the development of biomathematics as a subject in its own right is an answer to this question, fostering as it does the activities of teaching, consultation, and research at the interface between the mathematical and the biological sciences. Specialists of different kinds are drawn closer together, learn from each other, and begin to work in a more multidisciplinary environment (see, for instance, Sections 3.6 and 7.4). This has often proved to be highly advantageous in other areas of interest which have grown up initially at the periphery of several different well-defined disciplines. The spectacular development of modern genetics in recent years is a case in point.

In many, if not most, universities the primary concern is the teaching of undergraduates, though the claims of research and post-doctoral students have been receiving growing attention in recent years. From the community's point of view it may well be that the training of a new generation of teachers and specialists is the first requirement. Nevertheless, the promotion of research is also vitally necessary. It should be integrated with the other activities, rather than be regarded almost as a leisure option. When teaching programs become excessively heavy, any research interests the teachers have are the first to be sacrificed. Research then tends to become one's own private concern, a subject for polite small talk rather than serious cooperative endeavour. This stifles research, depresses teaching, and may be disastrous for the community.

Of course it is widely recognized that special efforts have to be made to preserve and encourage scientific research, and it is quite common to find in universities research organizations whose primary function is to prosecute research projects. While they may do some teaching or consulting work, they are to a large extent protected against serious encroachment by these routine activities. Unfortunately, when finance, accommodation, and equipment are in short supply it is more difficult for newer emerging disciplines to establish themselves. Biomathematics is such a discipline. Although various kinds of biomathematical work have been in existence for quite a long time, it is only fairly recently that the subject could begin to be recognized as a well-defined sphere of closely related studies. Far more institutions are accordingly required for the advancement of biomathematical science, and the organizational aspect is discussed further in the next section.

In the meantime let us look in more detail at the implications for specific research undertakings. We have already stressed the importance of early collaboration between biomathematician and biologist (or clinician). But quite often the situation will demand far more than the friendly cooperation of two colleagues. Several experts may be required, and specialized equipment may have to be used. Suppose, for instance, that an intensive study is to be made of the ecology of two competing species of animals within a large but well-defined habitat. To begin with, more than one biological expert may be required to cover adequately such aspects as zoology, physiology, ecology, genetics, biochemistry, etc. What should be included will depend on circumstances, and how many research workers are needed must be decided partly by the variety of skills available, and partly by how much laboratory and field work may be involved. Again, the biomathematical aspects may include statistical surveys of the habitat, the development of adequate stochastic models to represent the population interactions (birth, death, migration, predation, etc.), as well as elaborate computational facilities. We may easily end up with a team of two or three biologists plus a statistician and a computer expert. If all these individuals are to work together in harmony and finally produce a joint report on their work, the whole project must be carefully discussed and planned in advance. This does not mean that it must be completely rigid in form. Some flexibility is essential, but the criteria of scientific validity must also be observed (including the requirements of statistical and computational authenticity).

So far we have been thinking mainly of cooperative work in investigations which are biological, or in medical research as such. But another area in which cooperative projects are inescapable, and usually much harder to institute *ab initio*, is the sphere of operational research. The general principles and techniques of this subject have already been discussed at some length in Chapter 7, and we saw how a considerable range of experts with quite different backgrounds are frequently involved in operational research investigations. Any organization by its very nature will usually cover quite a number of different activities, and to understand these activities fully may require several specialists for a start. The mounting of a research program then needs the collaboration of one or two individuals with special knowledge and experience of operational research. Statistical surveys, data processing, numerical analysis, computer facilities, are all likely to be involved. Moreover, the validation of any tentative findings will necessitate actual operational trials. Existing activities may have to be partially reorganized. A good deal of sympathy and tact may be required to gain the willing cooperation of those whose jobs are under scrutiny.

Let us consider for a moment the kind of studies examined in some detail in Section 7.6, where the design of hospitals was discussed (see also Section 7.4). Any investigation into ward layout automatically involves patients, doctors, nurses, and other auxiliary staff, and may easily bring into the picture individuals working in related special departments such as X-ray, pathology, hospital records, etc. At the same time, debates about ward design or types of equipment may require special architectural and engineering skill. This immediately raises the fundamental problem of how large an operational research team should be.

In general, it will be impossible to have every relevant aspect fully represented by the appropriate expert. Such an attempt would normally lead to a team so large that it would be impossible to organize and direct it successfully. Some kind of compromise is inevitable, in which the major aspects are covered by specially appointed individuals, each of whom has secondary responsibilities as well. Thus senior members of the Nuffield Provincial Hospital Trust's team working on hospital design consisted of an architect with engineering qualifications, a doctor, a nurse, a sociologist, an historian, a retired army officer with previous experience of operational research in the services, and a statistician working in a medical school. This basic team size of seven is somewhere near the optimum, though different numbers may be demanded by different circumstances. In the Nuffield Trust team, however, fewer basic specialists would have left major aspects inadequately covered, and a larger number would have made the group unmanageable.

To begin with, the members of such a team must observe and discuss existing situations until they have a sufficient amount of common ground to decide on a generally agreed program of research. As new facts emerge and new proposals are put forward, a constant interchange of ideas will be necessary if the team is to maintain a reasonable homogeneity of outlook and to progress in an integrated way. Mutual cooperation at all stages is the key to successful operational research.

It is easy to see that there are far greater problems in the general coordination of operational research than there are in cooperative research within a more limited scientific context. Not only is a larger number of individuals with more diverse backgrounds involved in the actual carrying out of a research program, but there are usually many more sources of possible subsequent criticism when results are published. Thus new proposals on ward design, or hospital design generally, must be acceptable to the medical and nursing professions, as well as being architecturally sound and scientifically valid, not to mention being satisfactory to the principal consumer, the patient.

Many types of operational research, such as those involved in industry, business, the armed services, etc., are of necessity carried out largely in the location of application. They may or may not be linked with university research. It is on the whole desirable that there should be some connexion, as this is beneficial to both sides. Thus in universities with specific departments of operational research it is possible for academic teachers and research workers to participate quite actively in industrial applications, either as consultants or as part-time members of research teams. Conversely, those employed on the industrial side can help the university with seminars or special courses of lectures. Essentially similar arrangements can also be made with biological applications in say agriculture, fisheries, or forestry, or with medical applications in hospital management or public health. However, if the university already has a biomathematical department it may be advisable for this group to be primarily responsible for the biological and medical operational research in collaboration with the main operational research department.

Cooperative work between specialists in different disciplines is quite clearly fundamental to most biomathematical studies, whether these are mainly applied science or operational research. This fact must be carefully considered when discussing the broad shape of biomathematical organization, which we shall proceed to do in the next section.

9.4 THE ORGANIZATION OF BIOMATHEMATICS

In this section we shall attempt to sketch the broad outline of any institution or organization which undertakes the promotion of the integrated approach to biomathematics that we have repeatedly recommended in this book. It would of course be quite wrong to make a dogmatic pronouncement. What is required in any particular situation depends on what already exists, and there may be several different ways of satisfying given demands. For instance, suppose that in a university environment we are planning a new department of biomathematics. How would this be affected by pre-existing departments of statistics, operational research, or computing science? In the view of the writer, the close interrelation of all the mathematical skills and techniques involved in biomathematics makes it highly desirable that all the main features should be adequately represented within a single organization. The presence of other relevant institutions might influence the level of representation regarded as adequate, but should not affect the actual existence of that representation. The implications of this view will emerge more clearly from the subsequent discussion.

Let us consider the senior staff of a reasonably well provided for department or institute of biomathematics. To begin with, the four major aspects of mathematics, statistics (including probability), operational research, and computing should all be covered by suitably qualified individuals, preferably of full professorial rank. We have shown the need for and importance of these aspects in earlier chapters. They are all interrelated and overlapping, but nevertheless have well-defined centres of interest. Thus under the heading of *mathematics* we should want to have a wide range of pure and applied mathematics, including advanced algebra, complex variable theory, ordinary and partial differential equations, integral equations, the functions of mathematical physics, approximations and asymptotics, etc. Under *statistics and probability* we should put all mathematical statistics, including distribution theory, estimation, inference, sampling, surveys, experimental design, stochastic processes, simulation, etc. *Operational research* would involve the practical application of mathematical theories and techniques such as queueing theory, decision theory, linear programming, simulation, operational gaming, critical path analysis, systems analysis and system dynamics, etc., as well as the planning and organizing of operational research projects. Finally, *computing* would cover numerical analysis, scientific computing, data processing, and the provision of proper computing facilities for teaching, research, service requirements, on-line studies, etc.

Under each of these headings we must consider what is needed to cope properly with the demands of teaching, consultation and research. In general terms, the professor at the head of each major aspect or division should be supported by one or two readers (i.e. associate professors in American terminology) and three or four lecturers (or assistant professors). This scale of staffing may seem over-ambitious at first sight. It might be useful therefore to reflect for a moment on the range of tasks that might be required. Thus in teaching, biomathematics courses might be needed at undergraduate, postgraduate and research levels. In principle it might be supposed that biological and medical undergraduates could all attend the same lectures, but in practice it is frequently found that the needs of diverse subjects like agriculture, forestry, biochemistry, zoology, medicine, etc., are so different that it is quite impossible to run joint biomathematics courses that are sufficiently integrated with the other scientific work. This integration is vital if biomathematics is to be seen as relevant and important. While genuine duplication of effort is wasteful and undesirable, it is well worth providing courses that are specially tailored to groups of students with markedly different needs. Further, mathematically more advanced lectures should be available for mathematical undergraduates, as well as for those taking postgraduate diplomas or embarking on research.

Again, in the sphere of consultation, a careful check of the numbers of research students and senior research workers in the complete range of biological and medical departments who might benefit from and request biomathematical assistance could easily lead to the conclusion that the above staff estimates were inadequate, rather than the reverse.

One cannot repeat too often that it is fatally easy for heavy teaching programs and routine consulting duties completely to overshadow genuine research. We emphasized in Section 9.2 the importance of the interrelations between teaching, consultation, and research, and recommended that everyone should have some experience of all three although the relative emphasis might vary considerably. One way to guard against heavy pressure from one area unduly affecting the others is to have staff members with primary responsibilities in the different areas. Thus, in statistics, we might have one lecturer principally concerned with lecturing and informal teaching, one lecturer mainly occupied with planning and providing consultant advice, and one lecturer chiefly carrying out research. (The use of the same title of 'lecturer' for all three might be kept quite deliberately to avoid the growth of too-rigid distinctions.) The readers in each division should have had a considerable experience of research, and would have special responsibility for encouraging, directing, and formulating research projects, while the professors would have general oversight of all activities in their special divisions.

The above twofold classification is suggested as a means of maintaining a reasonable balance between teaching, consultation, and research, as well as between mathematics, statistics, operational research, and computing. It is not intended as an inflexible blueprint, but as a working guide to be varied according to conditions.

Suppose for instance that the university in question already had a strong department of mathematical statistics. The biomathematics department would still need its statistics division, but this could then afford to have less theoretical coverage and be stronger on applications. Some of the more theoretical lectures would perhaps be given by the mathematical statisticians, and so on.

Again, if there were a department of operational research, the biomathematics department should have an operational research division of its own to meet special medical requirements, though the staffing could probably be on a reduced scale. Of course, if the university had no medical school or local hospital looking to the university for biomathematical expertise, the operational research division might be dispensed with altogether, but not otherwise.

Similarly, the existence of a university computing centre should not mean that the biomathematics department is not equipped with its own computing division. We have already argued in Chapter 6 that the influence of computers on research is so fundamental that there is a strong case for small independent computers for teaching, program development, small-scale research, research involving man–machine interactions, on-line work, especially in connexion with operational research trials, etc. The availability of a large computing centre would be of considerable value in handling many routine computations, large-scale research requirements, and in providing expert advice on the finer points of numerical analysis and data processing. The essential thing is that the biomathematics department should at all times have immediate access to adequate computing facilities.

A broadly-based biomathematics department, as described above, needs an administrative head. Whether this should be a director over and separate from the four professors in charge of the major divisions, or whether it should be a chairman (possibly rotating) chosen from amongst them is largely a matter of local policy. What is important is that there should be a clearly defined assignment of responsibility. While the primary purpose of a biomathematics department is the carrying out of biomathematical work, this inevitably entails a whole host of administrative activities, both internal and external. Internal administration involves the day-to-day running of the department as well as long-term planning. It is essential that the director be provided with sufficient administrative and secretarial staff. This is specially important if the director is in fact one of the professors acting as a managing chairman, otherwise his personal academic activities, particularly research, will be largely neutralized. Generally speaking, each professor should have the services of a full-time secretary, with additional typing assistance if his own division produces a normal flow of research papers, reports, books, etc. External administration involves relations with other departments, the central university authorities, and outside bodies concerned with practical applications or with the provision of special grants. All of this is a full-time job, and it will not be done properly unless a single individual has the time and facilities for it, and can moreover be seen to be responsible for what happens.

It follows that directors should be given a fairly free hand in the running of their departments. They must of course be ultimately answerable to their faculty board, management committee, central authority, etc., depending on how things are organized. The higher authority should not interfere in day-to-day affairs. Except in case of emergency, reporting back should occur no more frequently than once a year, perhaps less often. On the other hand a full review of the department's activities and achievements should be undertaken at least once every five years. This kind of freedom is necessary if a proper spirit of adventure and enterprise is not to be stifled by petty restrictions and regulations. On one side we must meet the exigencies of planning (discussed in Section 9.1) in regard to the internal work of the department, the general form of university policy, and the social needs of the community. On the other side, we want to foster and encourage individual initiative, without which valuable opportunities will be wasted.

We have suggested that the foregoing scale of senior staff appointments is necessary for the proper establishment of biomathematics in a modern university with a full range of biological and medical departments. However, in a small or new university a start could be made on a reduced scale with the four main divisions headed by readers with a single professor as director. But it would be important that the growth of the other scientific departments be marked by a corresponding expansion in biomathematics itself. Otherwise the latter would carry insufficient weight to maintain its status and would soon have insufficient staff to perform efficiently the tasks required of it.

We have been discussing the formation of a biomathematics organization within a frankly university context. For various reasons this is likely to be the preferred location, although the possibility of an independent research institute is one which should also be seriously considered. When a fully integrated system of teaching, consultation, and research is envisaged, close association with a university is almost inevitable for the teaching alone, at least when it concerns undergraduates. But if a high-powered attack were to be launched on a wide range of fundamental biological and clinical problems, the establishment of an independent research institute might well be preferred. Relative isolation from the distractions of university life, both at senior and junior levels, might be seen as a positive advantage. At the same time the maintenance of a close association with ongoing biological, medical, and public health research would still be a *sine qua non*, and the provision of facilities for university research students would be highly advantageous if not essential. Also, steps would have to be taken to ensure complete interchangeability of staff with the universities, otherwise first-class research workers would be unwilling to take the risk of entering an outside organization.

We shall not attempt to sketch in further details of a biomathematics department or institute as these would be best worked out by the senior staff themselves. At the same time it should perhaps be emphasized that the provision of adequate funds is a not unimportant factor. In a university context the academic posts would all be permanent university appointments guaranteed by the university itself. Many short-term projects, on the other hand, of say two to five years in duration could be supported by outside funds. This would mean that the permanent staff of the department would be augmented by young temporary research workers who would provide additional enthusiasm and inspiration as well as learning from their senior colleagues. Questions of finance and fund-raising pose many perplexing problems. It can easily happen that a harassed professorial head of department spends much of his time raising money in a series of small contributions to carry out research that he has no time actively to take part in. With a full-time director adequately supported by administrative helpers, the position is different. An important part of his job is fund-raising. He can plan and organize it, and give his full attention to it without feeling that his real work is being strangled.

Biomathematics is today a growing and essential adjunct to the further development of the biological, medical, and public health sciences. If it is to play its proper role in an efficient and inspiring way it must be planned and organized on a scale commensurate with the importance of its contributions. The discussion of the present section has indicated some of the ways in which this might be done.

9.5 SPECIAL PROBLEMS OF BIOMATHEMATICS IN HEALTH

In the previous sections of this chapter we have been considering some of the problems of organizing biomathematics, taken in a rather general way so as to include applications to biology, medicine, and public health. Certainly, from the point of view of a university biomathematics department or a special biomathematics research institute, broad coverage seems desirable. This not only encourages greater efficiency by permitting the development of approaches which can be used in different fields of application, but can also stimulate interest and motivation through the interaction between varied types of investigation. However, as we have already seen, special circumstances may necessitate concentration on some individual type of activity, e.g. teaching, consultation, or research. In a similar way, the highly specific requirements of medical schools, hospitals, or ministries of health may inevitably lead to a narrowing down of the total range of biomathematical potentiality to a much more limited area.

There is also another powerful reason for accepting such restrictions. It is clear, of course, that fundamental research work in biology or medicine is in general most appropriately pursued in the academic environment of a university or similar institution. A certain degree of distance from the every-day problems of practical applications can be highly conducive to original thought. This is still true of some aspects of basic medicine. But as soon as physicians, surgeons, nurses, pharmacists, health administrators, etc., have to deal with individual people or collections of people, whether for curative or preventive purposes, all kinds of immediate operational problems arise. These include diagnosis, prognosis, the calculation of risks for alternative treatments or interventions, the weighing up of conflicting psychological, social, economic, or political factors, and so on. We have already dealt with such matters at some length in earlier chapters, and it is evident that they cannot be adequately handled in an abstract, theoretical manner.

Most health applications of biomathematics must accordingly be very closely tied to practical work, whether at the level of direct care for individual patients, or at the more general levels of population health services and overall health planning. Even in a university setting medical statistics is liable to be seen as a special subject in its own right, closely linked to the activities of a nearby teaching hospital and to the work of local community health services. However, if a medical statistical group, though practically oriented, becomes too divorced from other university activities involving mathematics, computing, operational research, etc., there is liable to be an appreciable loss of technological power and efficiency. A more widely based biomathematical department, with clear responsibilities for supporting applied work in the health field, might have a much more effective impact.

If such problems can arise even in a university setting, they are liable to occur even more acutely at the level of the regional administration of health services, and *a fortiori* within Ministries of Health. The higher one goes in the administrative hierarchy, the more one is potentially concerned with the spheres of operational research, systems analysis, system dynamics, and policy analysis. The extent to which these disciplines are actually invoked varies considerably. In some countries operational research and systems analysis groups work very closely with top administrators and are directly involved in the decision-making process. In other countries this happens hardly at all.

The consequence of the wide difference in working environments between academic biomathematical workers and practical, decision-oriented analysts can involve a considerable loss to both theory and practice. In some cases, there are academics developing and teaching mathematical refinements of basic operational research techniques in universities, while those wrestling with practical problems in industry, commerce, or government are penalized by insufficient access to technical mathematical support.

These problems are not easy to solve. If people dealing with day-to-day decision making in a complex environment are offered elaborate, expensive, computerized and incomprehensible mathematical tools, such tools are liable to be rejected. But if simple operational approaches are proposed for practical use and are capable of being constantly submitted to the test of experience, they are very likely to be accepted as a feasible form of technological assistance. In such a situation it should be possible to maintain close contact with more theoretical groups in universities, on an exchange or consulting basis. With care this kind of arrangment can result in mutual enrichment. The academic theorists can be more oriented towards practical applications and have real opportunities to test their ideas in practice. At the same time, those who are committed to immediate advice and action can have their assumptions and models examined more thoroughly and tested by university-based researchers possessing the requisite opportunities and skills.

Obviously, the details of this cooperation will vary widely with circumstances. But it seems essential for there to be a much closer collaboration between health authorities, both at regional and governmental levels, and university research institutions with regard to a wide area of biostatistics, medical statistics, operational research, and systems analysis. If this integration is not achieved, both spheres of activity will be increasingly impoverished.

Similar problems occur on an even greater scale at the international, as opposed to the national, level. What is required is a clearer delineation of functions at the various levels, e.g. university, governmental, international, together with mechanisms for improved communication between levels of problems and their solutions. For example, the ultimate evaluation of an internationally sponsored, multi-country trial of a new combined vaccine against several infectious diseases might very well depend, *inter alia*, on a better understanding of the population dynamics of the diseases in question when they occurred simultaneously, with possible interactions, in given communities. In spite of an already large and growing literature on infectious disease theory, little exists at the time of writing on such multiple, interacting disease situations. The required mathematical research could hardly be carried out very effectively at an international level. On the other hand, it might be possible to formulate the practical problems in a mathematical way such that research workers at a number of universities could collaborate with the international authorities and thus investigate theoretical problems whose solutions could have immediate practical consequences.

This type of collaboration still remains to be fully exploited. But an improved understanding of common interests between such bodies as the International Institute of Statistics, the Biometric Society, the International Federation of Information Processing, the International Epidemiological Association, the World Health Organization, etc., together with professionals at both governmental and university levels, could do much to bring about a substantial increase in the practical value of mathematical approaches to major world health problems.

CHAPTER 10

Future Prospects

We have now completed our relatively brief survey of the role of mathematics, statistics, modelling, computers, operational research, systems analysis, and system dynamics in assisting developments in medicine and public health. As can be seen, this book has not been primarily about these applied mathematical disciplines as such, but rather about what a wide collection of mathematically based attitudes, approaches, and techniques can do for people and their health.

For this reason we started in the first chapter with an overall view of the terrifying and oppressive facts of universal disease, as compared with the ideals of perfect health and well-being. We then examined the support that could be given by quantitative methods in dealing with some of the major problems of medical science and public health control. A basically rational and scientific approach led automatically to an extended discussion of the range and scope of the applied mathematical disciplines referred to above. It has been a recurrent theme of our presentation that, though many of these disciplines do involve considerable technical sophistication for their professional handling, a general understanding of the basic ideas and their implications can be derived from straightforward commonsense notions available to almost everybody.

We have therefore attempted, for the most part, to restrict discussion to general principles and concepts, to avoid the presentation of purely technical matters, and to emphasize the practical consequences of adopting and applying the various types of mathematical, or quasi-mathematical, expertise available. Two obvious advantages follow from this treatment. First, it should enable the quantitatively oriented reader to have a better idea of what mathematics, statistics, and systems can offer in support of active work in the health field (one might also hope that the less mathematically inclined reader could gain a better idea of some of the existing achievements and new potentialities). Secondly, the treatment should facilitate better communication between those professionals who are primarily concerned with medicine and public health and the experts in the mathematical sciences who can provide valuable assistance in certain areas.

In a word, applied mathematics in a top-priority area like health is too important to be left *solely* in charge of mathematicians. (Of course, *mutatis mutandis*, the same thing could be said of medicine and health!) It is therefore essential that there is sufficient comprehension of each other's skills and capabilities by all professionals in this multidisciplinary situation. Individual technical responsibility must naturally remain with those who have the requisite qualifications and experience, whether we are talking of specific clinical subjects, epidemiological activities, public health control, or various technological aids such as can be provided from the mathematical sciences. This simply means that the mathematicians are responsible for the mathematics as such, i.e. the handling of the medical statistics, the manipulation of mathematical models, the control of computerized calculations, etc. But the choice of problems to be solved, the general terms in which the problems are couched, the broad structure of any models that are used, the interpretation of results obtained, the validation of hypotheses against appropriate factual observation, and so on, are all matters requiring understanding and agreement on a multidisciplinary level between a variety of experts with different backgrounds.

We must also be careful not to lose sight of the ultimate customers—suffering humanity. Doctors should not, for example, uncritically accept the high-pressure publicity of pharmaceutical companies. Nor should they believe the alleged implications of some esoteric mathematical calculation or the facile conclusions of a simplified computer printout without asking why they should believe those things, what principles were involved, and what kind of assumptions were made. Conversely, mathematicians, statisticians, systems analysts, and computer experts should not in general agree to handle problems formulated by those working in other disciplines without understanding enough of the practical aspects to judge whether the problems have been posed in a sufficiently precise and relevant way. Similar considerations apply to the general public who expect their diseases to be treated, cured, or prevented, and who want their broad standards of health and well-being to be promoted by the society they live in.

It is becoming increasingly common for people to carry on a useful dialogue with, say, an accountant handling a complicated income-tax return, or a garage mechanic reporting on the functioning of a motor-car engine. Similarly, more and more people discuss their medical and health problems in detail with the doctors concerned. They are, for the most part, not medically expert themselves. But they want to know, in commonsense terms, what is wrong, what will happen, what treatment is proposed, how long it will take, what it will cost, what are the risks involved, and so on. They may even sue their doctors if they are dissatisfied with the results!

Historically speaking, this is an excellent trend, but if the customers, i.e. the patients, are really to get the best help from their technical advisers, i.e. the doctors and public health authorities, there must be more mutual trust, understanding, and cooperative action. It is widely appreciated that patients should not be kept in ignorance of what is to be done to them, whether in a general practitoner's surgery or in a hospital ward or clinic. Abuses do occur, but the basis for improved cooperation is already understood. More difficult is a general appreciation of the practical implications of the specialized matters discussed in the present book.

We have argued that simple arithmetical ideas arise in commonsense approaches to everyday problems, and that elementary skills in assessing odds and probabilities are also widely distributed. Such capabilities are the indispensable basis for a wider, sympathetic but critical, contact between the general population and medical or health authorities. But more needs to be done when the activities involved are on a vast scale, involving a complex system of processes handled by intricate computerized methods. As yet, no one is very good at these things, in the sense of being able to provide rational methods of controlling large-scale social processes in such a way a to command general informed acceptance.

Naturally, all the topics discussed in this book require further development, amplification, and simplification, so as to make them more readily accessible not only to professionals in the health area who could use them, but also to the lay public who are interested in the final consequences for their daily lives.

In making an assessment of future prospects for mathematics, statistics, and systems in medicine and health we are specially concerned with what ought to happen, and what we should make special efforts to bring about. Of course, it goes without saying that such matters are highly speculative, and that there is as yet no clear central mechanism for the control of policy and development in health mathematics—and many would argue that such control is in fact undesirable. However, certain trends can be discerned. And, in so far as these are more widely understood and appreciated, future discussion of available strategic options will be facilitated.

We have been occupied throughout with the *application* of various mathematical sciences to medical and health problems, rather than with the purely technical aspects internal to those sciences. From this point of view, therefore, how do we stand with regard to present trends? As we have already noted, there is a strong tendency for some mathematicians to seize on a practical problem, simplify it down to the point where it is mathematically tractable, and then develop solutions in great detail that are of primary interest only to certain other mathematicians. Sometimes a so-called practical illustration is tacked on at the end to show a tenuous connexion with reality. This is the main danger to be avoided by mathematical modelling if it is to retain practical importance. Although this tendency does little direct harm—most of the mathematical work is simply ignored, except by a small number of mathematical experts—it does indirectly considerably reduce the amount of mathematical support available to practically oriented health undertakings.

The trend could probably be overcome by increased efforts, both at national and international levels, to employ more mathematicians on applied problems, so as to make more opportunities available for them to work directly with medical scientists in laboratories and hospitals, and with public health authorities on the wider problems of intervention, prevention, and control. Such opportunities would provide difficult, rather than easy, problems for solution. The requirements to contribute to real scientific or administrative work would also make the publication of elegant mathematical papers in technical journals more difficult. On the other hand, the challenge might well produce a surprising response. New forms of socially oriented applied mathematics of real practical importance might arise. Moreover, the constant need to maintain existing professional standards would entail much patient mathematical work of a familiar kind—the difference being that the latter would involve studies of rigour, the proving of theorems, and other mathematical niceties, not just for their own sake, but because they were part of the process of underpinning and guaranteeing the reliability of the overall support provided.

Another trend we have already noted is the tendency of certain kinds of medical and health statistics to accumulate vast mountains of data that go far beyond any kind of cost–benefit justification. The tremendous advance in recent years in the capabilities and availability of computerized data-handling facilities has intensified this trend. Many project directors seem to be quite happy to supplement their descriptive reports of data collected with hundreds of computer-produced tables, which produce few reliable conclusions as they stand, and which most people do not read anyway. We have mentioned that the advent of cheap, pocket-sized electronic computers is enabling mathematicians and statisticians to escape from the tyranny of the big computer installations. At the same time, the latter are unbeatable when it comes to coping with large data bases, intricate multivariate studies, or complex system dynamical investigations.

A new look is therefore required at any massive, routine statistical operation. Are the data really needed, at least on the scale envisaged? Who uses them? How often, and what for? What is the approximate cost–benefit ratio? Would it not be better to mount far smaller, *ad hoc* studies, geared to specific decision-making activities? And so on. Such questions must be asked more clearly, and answered. Every effort must be made to streamline those statistical operations so that newer methods more adapted to dealing with large-scale activities can be taken advantage of.

Generally speaking, the problems of future mathematical developments are probably least in the area of gaining fundamental knowledge in basic biology and clinical medicine. Here the standard methods of scientific investigation apply, and the use of quantitative methods can be traced back for centuries. True, it would help if more opportunities for applied mathematical work were made available, but no revolutionary change in traditional methods of operation seems to be required. Most of the work takes place in the research laboratories attached to universities, hospitals and other biological and medical research institutions. This view does not of course deny that, from time to time, general policy decisions may have to be taken on which lines of investigation show most promise and what should be the relative intensities of financial support.

The most controversial area at the present time is the one covering the more ambitious kinds of operational research, namely the large-scale use of macro-modelling systems analysis and system dynamics. People are rightly afraid of being inveigled into accepting the dictates of a new-fangled technology involving the application of prestigious and expensive gadgetry that they cannot understand or rationally evaluate. This is quite reasonable. But, on the other hand, complex, interacting, and counter-productive social systems exist in real life for all to see. Administrative authorities act as though they believe that they understand how to manipulate the real-life systems effectively. Most of us have serious doubts about the truth of those beliefs.

Progress can accordingly be made only if we can develop methods of handling large-scale phenomena more surely, whether we are dealing with agriculture, economics, education, environmental hazards and pollution, industrial production, or health. Indeed, all these sectors, and many others, strongly interact with each other. It is not so much a question of whether we shall adopt a systems approach, but what kind of systems approach is needed. Whatever can be said against existing forms of systems analysis, industrial dynamics, urban dynamics, the Club of Rome reports, health service modelling, econometric macro-modelling, etc., the fact remains that medical and health authorities, as well as the general public, must find new ways of co-operating and meeting the challenge imposed by an interlocking complex of physical, biological, and social processes, if the basic health needs of humanity are to be met in the future in a more imaginative, sympathetic, and efficacious manner.

The pursuit of health is thus an essentially cooperative endeavour, requiring intelligent and sympathetic collaboration between members of the general public and their technical advisers. The latter cover not only health professionals of all kinds, but also a wide range of scientific workers, including those who provide support from a variety of applied mathematical disciplines. If these mathematical skills are to contribute their full potential for health improvement in the future they must be far more integrated than at present with the relevant scientific, medical, epidemiological, and public health activities, and must in addition be more closely assimilated to commonsense methods of enquiry and evaluation. Only in this way can the elegant and austere abstractions of mathematics make a positive contribution to the fight against disease and the preservation of human health and well-being.

REFERENCES

Anderson, J. A. and Boyle, J. A. (1968). Computer diagnosis: statistical aspects. *Brit. Med. Bull.*, **24**, 230.

Armitage, P. (1960). *Sequential Medical Trials*, Blackwell, Oxford.

Ashby, R. W. (1952). *Design for a Brain*, Wiley, New York.

Bailey, N. T. J. (1952). A study of queues and appointment systems in hospital outpatient departments, with special reference to waiting-times. *J. Roy. Statist. Soc.*, B, **14**, 185.

Bailey, N. T. J. (1954). Queueing for medical care. *Applied Statistics*, **3**, 137.

Bailey, N. T. J. (1956). Statistics in hospital planning and design. *Applied Statistics*, **5**, 146.

Bailey, N. T. J. (1962). Calculating the scale of inpatient accommodation. In *Towards a Measure of Medical Care*, Oxford University Press, London, p. 55.

Bailey, N. T. J. (1964). *The Elements of Stochastic Processes*, Wiley, London.

Bailey, N. T. J. (1975). *The Mathematical Theory of Infectious Diseases*, Griffin, London.

Bailey, N. T. J. and Thompson, M. (Ed.) (1975). *Systems Aspects of Health Planning*, North-Holland, Amsterdam.

Bartlett, M. S. (1966). *An Introduction to Stochastic Processes*, 2nd ed. Cambridge University Press, London.

Bharucha-Reid, A. T. (1960). *The Elements of the Theory of Markov Processes and their Applications*, McGraw-Hill, London.

Churchman, C. W., Ackoff, R. L., and Arnoff, E. L. (1957). *Introduction to Operations Research*, Wiley, New York.

Cochran, W. G. and Cox, G. M. (1957). *Experimental Designs*, 2nd ed. Wiley, London.

Copernicus, N. (1543). *De revolutionibus orbium coelestium*, Nuremberg. First complete English translation in *Great Books of the Western World*, 1952, Vol. 16. Encyclopedia Britannica Educational Corp., Chicago.

Cox, D. R. and Miller, H. D. (1965). *The Theory of Stochastic Processes*, Wiley, New York.

Cox, D. R. and Smith, W. L. (1961). *Queues*, Methuen, London.

Crooks, J., Murray, I. P. C., and Wayne, E. J. (1959). Statistical methods applied to the clinical diagnosis of thyrotoxicosis. *Quart. J. Med.*, **28**, 211.

Davies, P. (1972). Symptom diagnosis using Bahadur's distribution. *Bio-Med. Comp.*, **3**, 307.

Enslein, K. (Ed.) (1962). *Data Acquisition and Processing in Biology and Medicine*, Pergamon, Oxford.

Enslein, K. (Ed.) (1964a). *Data Acquisition and Processing in Biology and Medicine*, Vol. 2, Pergamon, Oxford.

Enslein, K. (Ed.) (1964b). *Data Acquisition and Processing in Biology and Medicine*, Vol. 3, Pergamon, Oxford.

Enslein, K. (Ed.) (1966). *Data Acquisition and Processing in Biology and Medicine*, Vol. 4, Pergamon, Oxford.

Enslein, K. (Ed.) (1968). *Data Acquisition and Processing in Biology and Medicine*, Vol. 5, Pergamon, Oxford.

Feller, W. (1957). *An Introduction to Probability Theory and its Applications*, Vol. 1, 2nd ed. Wiley, New York.

Fisher, R. A. (1925). *Statistical Methods for Research Workers*, Oliver and Boyd, London.

Fisher, R. A. (1935). *The Design of Experiments*, Oliver and Boyd, London.

Flagle, C. D., Huggins, W. H., and Roy, R. H. (1960). *Operations Research and Systems Engineering*, Johns Hopkins, Baltimore.

Forrester, J. W. (1961). *Industrial Dynamics*, MIT, Cambridge, Mass.

Forrester, J. W. (1969). *Urban Dynamics*, MIT, Cambridge, Mass.

Forrester, J. W. (1971). *World Dynamics*, Wright-Allen, Cambridge, Mass.

Forsyth, G. and Logan, R. F. L. (1960). *The Demand for Medical Care*, Oxford University Press, London.

Galton, F. (1889). *Natural Inheritance*, Macmillan, London.

Gardner, H. (1973). *The Quest for Mind*, Knopf, New York.

Garvin, W. W. (1960). *Introduction to Linear Programming*, McGraw-Hill, London.

Gurland, J. (Ed.) (1964). *Stochastic Models in Medicine and Biology*, University of Wisconsin, Madison.

Hamming, R. W. (1962). *Numerical Methods for Scientists and Engineers*, McGraw-Hill, London.

Hippocrates. *Airs, Waters and Places*. English translation: Loeb Classical Library, London, 1923–31, Hippocrates, Vol. 1.

Hollingdale, S. H. and Tootill, G. C. (1975). *Electronic Computers*, 2nd ed. Penguin, Harmondsworth, England.

Hoos, Ida R. (1969). *Systems Analysis in Social Policy*, Institute of Economic Affairs, London.

International Institute for Applied Systems Analysis (1974). *Background Information*, IIASA, Laxenburg, Austria.

Illich, I. (1975). *Medical Nemesis*, Calder and Boyars, London.

Jacquez, J. A. (Ed.) (1964). *The Diagnostic Process*, University of Michigan, Michigan.

Jeffreys, H. (1961). *Theory of Probability*, 3rd ed. Clarendon, Oxford.

Kendall, M. G. and Stuart, A. (1969 and later). *The Advanced Theory of Statistics*, Vols. 1–3, Griffin, London.

Koestler, A. (1959). *The Sleepwalkers*, Penguin, Harmondsworth, England.

Lalonde, M. (1974). *A New Perspective on the Health of Canadians*, Information Canada, Ottawa.

Ledley, R. S. and Lusted, E. B. (1961). Mathematical diagnosis and modern decision making. In *Mathematical Problems in the Biological Sciences* (Ed. R. Bellman), Proceedings of a Symposium in Applied Mathematics, Vol. 14, American Math. Soc., Providence, RI, p. 117.

Lindley, D. V. (1965). *Introduction to Probability and Statistics from a Bayesian Viewpoint*, Vols. 1–2, Cambridge University Press, London.

Lotka, A. J. (1925). *Elements of Physical Biology*, Williams and Wilkins, Baltimore. Subsequently reissued in 1956 as *Elements of Mathematical Biology*, Dover, New York.

Majone, G. (1975). The feasibility of social policies. *Policy Science*, **6**, 49.

Masuyama, M. (1963). Various approaches to automatic diagnosis by discrete variables. *Dep. Statist. Appl. Res. Un. Jap. Sci. Engrs.*, **10**, 151.

Meadows, D. H., Meadows, D. L., Randers, J., and Behrens, W. W. (1972). *The Limits to Growth*, Universe Books, New York.

Mesarovic, M. and Pestel, E. (1974). *Mankind at the Turning Point*, Hutchinson, London.

Molina, E. C. (1942). *Poisson's Exponential Binomial Limit*, Van Nostrand, New York.

Neumann, J. von, and Morgenstern, O. (1947). *Theory of Games and Economic Behavior*, Princeton.

Norusis, M. J. and Jacquez, J. A. (1975). Diagnosis. I. Symptom nonindependence in mathematical models for diagnosis. *Comp. Biomed. Res.*, **8**, 156.

Nuffield Provincial Hospitals Trust (1955). *Studies in the Function and Design of Hospitals*, Oxford University Press, London.

Nuffield Provincial Hospitals Trust (1965). *Waiting in Outpatient Departments*, Oxford University Press, London.

Parzen, E. (1962). *Stochastic Processes*, Holden-Day, San Francisco.

Payne, L. C. (1963). Towards medical automation. *Wld Med. Electronics*, **2**, 7.

Pearson, K. (1892). *The Grammar of Science*, Everyman Edition, 1938.

Quetelet, L. A. J. (1835). *Sur l'homme et le développement de ses facultés, un essai de physique sociale*, Bacheliér, Paris.

Rashevsky, N. (1960). *Mathematical Biophysics*, Vols. 1–2, 3rd ed. Dover, New York.

Rashevsky, N. (1961). *Mathematical Principles in Biology*, Thomas, Springfield, Illinois.

Saaty, T. L. (1961). *Elements of Queueing Theory*, McGraw-Hill, London.

Sasieni, M., Yaspan, A., and Friedman, L. (1959). *Operations Research*, Wiley, London.

Scheinok, P. (1972). Symptom diagnosis: Bayes's theorem and Bahadur's distribution. *Bio-Med. Comp.*, **3**, 17.

Shuman, L. J., Wolfe, H., and Speas, R. D. (1974). The role of operational research in regional health planning. *Operations Res.*, **22**, 234.

Takahashi, K., Dohi, Y., Sato, T., Hasegawa, T., Ōuchi, A., Takahashi, T., and Fukumachi, M. (1963). Logic of medical diagnosis and a matrix-type electric diagnostic machine. *5th Int. Conf. Med. Electronics*, Liège.

Thompson, D'A. W. (1st ed. 1917; 2nd ed. 1942). *On Growth and Form*, Vols. 1–2, Cambridge University Press, London.

Vishnevsky, A. A., Artobolevsky, I. I., and Bykhovsky, M. L. (1964). Cybernetic methods in medicine. *Scientific World*, **8**, 13.

Warner, H. R., Toronto, A. F., Veasey, L. G., and Stephenson, R. (1961). A mathematical approach to medical diagnosis. *J. Amer. Med. Assoc.*, **177**, 177.

Wiener, N. (1948). *Cybernetics*, Wiley, London.

Wittgenstein, L. (1921). Logisch–Philosophische Abhandlung, *Annalen der Naturphilosophie*, XIV, 185–262. Published in 1922 as *Tractatus Logico–Philosophicus* with English translation, Routledge and Kegan Paul, London.

Woerkom, A. J. van and Brodman, K. (1961). Statistics for a diagnostic model. *Biometrics*, **17**, 299.

Index of Authors

Index of Subjects

Applied Probability and Statistics (Continued)

GOLDBERGER · Econometric Theory

GROSS and CLARK · Survival Distributions: Reliability Applications in the Biomedical Sciences

GROSS and HARRIS · Fundamentals of Queueing Theory

GUTTMAN, WILKS and HUNTER · Introductory Engineering Statistics, *Second Edition*

HAHN and SHAPIRO · Statistical Models in Engineering

HALD · Statistical Tables and Formulas

HALD · Statistical Theory with Engineering Applications

HARTIGAN · Clustering Algorithms

HILDEBRAND, LAING and ROSENTHAL · Prediction Analysis of Cross Classifications

HOEL · Elementary Statistics, *Fourth Edition*

HOLLANDER and WOLFE · Nonparametric Statistical Methods

HUANG · Regression and Econometric Methods

JAGERS · Branching Processes with Biological Applications

JESSEN · Statistical Survey Techniques

JOHNSON and KOTZ · Distributions in Statistics
Discrete Distributions
Continuous Univariate Distributions-1
Continuous Univariate Distributions-2
Continuous Multivariate Distributions

JOHNSON and KOTZ · Urn Models and Their Application: An Approach to Modern Discrete Probability Theory

JOHNSON and LEONE · Statistics and Experimental Design: In Engineering and the Physical Sciences, Volumes I and II, *Second Edition*

KEENEY and RAIFFA · Decisions with Multiple Objectives

LANCASTER · An Introduction to Medical Statistics

LEAMER · Specification Searches: Ad Hoc Inference with Nonexperimental Data

LEWIS · Stochastic Point Processes

McNEIL · Interactive Data Analysis

MANN, SCHAFER and SINGPURWALLA · Methods for Statistical Analysis of Reliability and Life Data

MEYER · Data Analysis for Scientists and Engineers

OTNES and ENOCHSON · Digital Time Series Analysis

PRENTER · Splines and Variational Methods

RAO and MITRA · Generalized Inverse of Matrices and Its Applications

SARD and WEINTRAUB · A Book of Splines

SEARLE · Linear Models

THOMAS · An Introduction to Applied Probability and Random Processes

WHITTLE · Optimization under Constraints

WILLIAMS · A Sampler on Sampling

WONNACOTT and WONNACOTT · Econometrics

WONNACOTT and WONNACOTT · Introductory Statistics, *Third Edition*

WONNACOTT and WONNACOTT · Introductory Statistics for Business and Economics, *Second Edition*

YOUDEN · Statistical Methods for Chemists

ZELLNER · An Introduction to Bayesian Inference in Econometrics